Cosmos and Anthropos

Cosmos and Anthropos

A Philosophical Interpretation of the Anthropic Cosmological Principle

Errol E. Harris

Humanities Press International, Inc.
New Jersey ◇ London

First published 1991 by Humanities Press International, Inc.,
Atlantic Highlands, N. J., and 3 Henrietta Street, London WC2E 8LU

© Errol E. Harris, 1991

Library of Congress Cataloging-in-Publication Data

Harris, Errol E.
    Cosmos and anthropos: a philosophical interpretation of the
anthropic cosmological principle / Errol E. Harris.
        p.   cm.
    Includes bibliographical references and index.
    ISBN 0–391–03694–7
    1. Cosmology. I. Title. II. Title: Anthropic cosmological principle.
BD701.H37   1991
113--dc20   90–37036
                                                                  CIP

British Cataloging in Publication Data

A CIP record for this book is available from the British Library.

Printed in the United States of America

Contents

Preface

In October 1986, I attended a symposium at Colorado State University on the question: Does the New Physics need a new Metaphysics? Six physicists and three philosophers took part in the discussion. I was delighted, although not entirely surprised, to find that the scientists, virtually unanimously, called for the kind of holistic metaphysic (with the accompanying logic of internal relations) that I had been advocating and trying to elaborate for more than thirty years—in fact throughout my philosophical career. Then, two years later, I was invited to take part in a physics workshop at George Mason University, on "Bell's Theorem and the Conception of the Universe." The paper I offered, on the Anthropic Principle, was so well received by the assembled physicists, and the general enthusiasm for ideas I had so long espoused was so marked, that I was persuaded, against earlier intentions, to write another book. That was the stimulus and the occasion for what appears in the following pages.

This book is not a scientific treatise. I am neither mathematician nor physicist and have no qualification to attempt any such thing. All the scientific material here presented is borrowed from the writings of those who are competent to pronounce on such matters, and who, incidentally, are not responsible for any errors I may have committed. I have sought only to use this material as evidence in support of a philosophical conception of the world that seems to me to be the most convincing in the light of scientific findings. It is a philosophical position I have long since contended was implied by the 20th-century scientific revolution, but which has only very recently dawned on the scientists themselves, although, for many decades, there have been eminent thinkers among them who have suggested developments in the direction here taken.

Among philosophers in the present age, the only ones, of whom I am aware, who have seriously entertained a position of the kind suggested in this book are now largely overlooked or considered old-fashioned (except by a select group of followers and admirers). They are thinkers like Samuel Alexander and Alfred North Whitehead; while the work of R. G. Collingwood, who initiated many of the key ideas, in his book *An Essay on Philosophical Method*, is almost entirely neglected and forgotten.

Most contemporary philosophers have been, and still are, it seems, unimpressed by the massive evidence that scientists have produced, and are disinclined to consider its consequences for logic and metaphysics. This, no doubt, is the result of the widespread, long, and exclusive influence of the Analytic philosophy stemming from the interdict against metaphysics pronounced by the Logical Positivism that prevailed in the late 1930s, which banned philosophical thinkers from addressing metaphysical questions. It has dominated philosophical education (at least in the English-speaking world) ever since, and has prevented many of the most gifted minds among the younger generation from noticing the powerful (one might say, overwhelming) trends in contemporary physical science, and from trying to develop the sort of metaphysical position that they so copiously imply.

Being no scientist myself, I appeal to the reader not to judge this book simply as a scientific report, but to consider it as an attempt at metaphysical reconstruction, proceeding from the world view that contemporary science has generated. It must, of course, be conceded that not all the scientific theories to which I have appealed are yet fully established. Professor Roger Penrose ranks some of them as only tentative. Yet they are being developed, and the ideas contained in them are being pursued, by many of the foremost physicists and mathematicians of the day. One must also recognize that even the best established theories are liable to revision in the light of further discoveries. But this has been the case throughout the history of science, and philosophy has always been, throughout that same history, inseparably linked with scientific thinking—for much of the time virtually indistinguishable from it. Only when the two disciplines advance hand in hand are they likely to produce lasting results. Neither can afford to ignore the other, and even when the practitioners of one seem to despise and belittle the other, they are seldom free from mutual influence.

In the history of Western thought, from the ancient Greeks on, the greatest philosophers have themselves been among the most prominent scientists of the day (even when they were not so called). In modern history, Descartes, Leibniz, and Kant are eminent examples. Others, such as Schelling and Hegel, were deeply versed in the scientific literature of their period, and in the earlier years of the present century, Bergson, Whitehead, and Merleau-Ponty were notable cases in point. Unfortunately, contemporary science has become so technical and its scope has become so vast, that it is difficult, if at all possible, for the professional philosopher to be expert even in one of the main fields, let alone all of them; and the need for specialization makes it equally difficult for scientists to become sufficiently versed in the relevant philosophical literature.

Consequently, the more speculatively inclined scientists sometimes tend to state their philosophical convictions in ways that raise eyebrows among professional philosophers, and excite criticism that proves, in the final issue, to be merely superficial. This was the fate of Eddington and Jeans, some sixty years ago, at the hands of Susan Stebbing and C. E. M. Joad. Recent statements of the Anthropic Cosmological Principle have received a similar critical reception. What I have tried to do is to draw attention to the shortcomings of the statements, while insisting upon and trying to develop the more profound significance of the idea. However feeble my own attempts have been to achieve *rapprochement* between the two perspectives, I have not been able to resist the urge to continue.

My thanks are due to the Center for Philosophy and History of Science at Boston University for the facilities made available to me as an honorary research fellow, and especially to the Director of the Center, Professor Robert Cohen, for reading some of my manuscript and giving me the benefit of his comments. Again, he must not be held responsible for any mistakes that may persist in what I have written. I am extremely grateful to Professor Philip Grier for his unfailing help with technical difficulties in word-processing and printing out the manuscript. Finally, I wish to thank Professor Menas Kafartos and the Smithsonian Institute for inviting me to participate in the course on Design in the Universe, at which I first presented what now appears as Chapter 12 of this book.

E. E. H.
High Wray,
March 1990

1
The Anthropic Principle

STATEMENT AND CRITICISM

In recent years, physicists have suddenly discovered what they have called the Anthropic Principle, a conception that many present-day physicists regard not just as a speculative idea but as a serious scientific principle which, if respected, can give rise to significant observational predictions crucial to the acceptance of cosmological hypotheses. In view of attitudes that have prevailed in modern science virtually since its inception, the new development is quite surprising and will be viewed by many scientists and a large body of philosophers with scepticism and distaste, or even with dismay. The Anthropic Principle, however, ought not to be treated lightly nor should it be despised, for it points to an important truth that contemporary scientific theories, more than any in the past, have brought to the fore. The way it has been expounded by some of its proponents leaves much to be desired by way of coherence, and the forms in which it has been enunciated may well have objectionable implications, but the facts and considerations that support it have philosophical implications of the greatest importance which, in the light of what the sciences, from physics to sociology, have disclosed, are difficult to gainsay. We may begin by examining its shortcomings and then, allowing for these, go on to consider how it could be vindicated.

The principle has been stated in several forms, styled:

(1) the Weak Anthropic Principle (WAP): "The observed values of all physical and cosmological quantities are not equally probable but they take on values restricted by the requirement that there exist sites where carbon-based life can evolve and the requirement that the Universe be old enough for it to have already done so."[1]
(2) the Strong Anthropic Principle (SAP): "The Universe must have those properties which allow life to develop within it at some stage in its history."[2]
(3) the Participatory Anthropic Principle (PAP): "Observers are necessary to bring the Universe into being."[3]
(4) the Final Anthropic Principle (FAP): "Intelligent information-processing

1

must come into existence in the Universe, and, once it comes into existence, it will never die out."[4]

One would think it obvious that, if such a principle can be stated at all, there must be an intelligent being to state it, so that its being entertained by physicists is insurmountable evidence that they exist, and it follows immediately that the conditions necessary for their existence must therefore obtain, and as they are presumably examples of intelligent life inhabiting a part of the universe, the universe must be such, and so must be observed by intelligent observers to be such, as to provide these conditions. Why then should physicists only now find it necessary to propound the Anthropic Principle?

Of the principle in its first form Professor Stephen Hawking writes: "The weak anthropic principle states that in a universe that is large or infinite in space and/or time, the conditions necessary for the development of intelligent life will be met only in certain regions that are limited in space and time. The intelligent beings in these regions should therefore not be surprised if they observe that their locality in the universe satisfies the conditions that are necessary for their existence. It is a bit like a rich person living in a wealthy neighbourhood not seeing any poverty."[5]

What is it, one may ask, that leads Professor Hawking, or anybody else, for that matter, to expect that intelligent beings would be surprised by any such observation? The reason is that scientists until very recently have thought, and, as remarks like Professor Hawking's show, inadvertently still do think, of observation as something that impinges on the physical world, as it were, from the outside, without interfering with it; and of the world itself as a vast automaton that runs according to its own intrinsic laws, without relation to observers.

This is the inheritance passed down from the Copernican revolution at the time of the Renaissance and its consolidation in the Newtonian system of celestial and terrestrial mechanics. There is no record, so far as I know, of any ancient or mediaeval thinker expressing surprise that the universe should be such as to contain human beings; nor is it likely that they should, because the ancients regarded the cosmos as a living creature with an all-pervasive soul, in which human souls were individual participants; and the mediaevals inherited this conception with the qualification that the world was created by God for the sake of human-kind and His own glory. But the effect of the Copernican revolution, as we are so often told, was to remove the earth and man from the centre of the universe and to depict the world as a machine, created, no doubt, by God, but with its own laws which kept it working without further divine intervention. It was a machine which had no place in it for the human mind that came to know it, and its laws could not account for the nature of that mind, nor for its existence, nor for its relation to the human body which it inhabited and which was a part of the physical system. Hence Descartes considered the body to be a mere automaton, affected inexplicably by the mind, its relation to which was a perpetual problem to him and his successors. Mind and body, Descartes decided, belonged to two separate substances, which had nothing in common except their creator, God.

In these circumstances it would indeed have been surprising if human beings had found the physical world to be such as to provide the conditions necessary for the existence within it of minds. In fact, viewing the world as a machine which made no provision whatsoever for intelligent observers, so that their existence and consciousness were an impenetrable mystery, they could hardly have expected to discover in the celestial mechanics the conditions necessary for their own awareness, unless, like Hobbes, they reduced that awareness to just another example of matter in motion. But then, it still remained impenetrably mysterious how matter in motion could observe itself and other moving bodies.

For Descartes, unable to deny his own immediate awareness of his own existence, the problem was how he could become apprised of the machine-world around him, the reality of which was not so directly revealed. His empiricist contemporaries and successors were content to rely on the deliverances of their bodily senses, which Descartes distrusted. Through the mediation of the senses, they held, the "external world" could be observed, as if through a window (Locke's metaphor), or as it were from afar, through a telescope; and, for accuracy, care had to be taken that what was disclosed was not contaminated by any subjective influences. What was discovered of the universe was meticulously segregated from what the mind revealed of itself. For the world to afford the peculiar conditions necessary for the production within it of observing minds required a special dispensation from God. Mind was a separate creation, not of this world but belonging more appropriately to the Kingdom of Heaven.

These were the metaphysical presuppositions of science in the 17th and succeeding centuries, but the advent of Darwin and the theory of evolution in the mid-19th century introduced a new and revolutionary factor. Human beings and their mental capacities were now seen to have evolved from the non-human, and a bridge between matter and mind began to be conceivable, especially when, in the course of time, life came to be regarded as having developed from the non-living. The full effect of Darwin's theory was not at first appreciated, because the "mechanism" of evolution was for some time (and still is) assimilated to that of the physical world through chance variation and natural selection. But the 20th-century revolution in physics has changed everything. The universe is no longer conceived as a machine. Life can now be more easily understood as a development continuous with the non-living; and the world is observed as providing conditions for the emergence of intelligent beings.

For those thinkers still under the spell of the Newtonian world-view, this will naturally be surprising, despite the obvious fact that for intelligent life to exist in the world the conditions requisite for its emergence within physical nature must necessarily obtain. It is therefore natural for Professor Hawking to expect surprise, and for physicists now to question that surprise and to enunciate the Anthropic Principle. But if the principle is so obvious as to be a virtual truism, what is its real significance?

To deny the existence, or even to assume the non-existence, of intelligent life is rather like denying (or assuming the falsity of) *Cogito ergo sum*. The assertion immediately refutes itself. Nothing could be more obvious than that, were there no intelligent life in the universe, there would be neither observers nor scientists to pronounce anthropic principles, nor any who might question their validity. In short, this discussion would not be taking place and would not be possible. To speculate on what the universe might be like if it were unobserved is akin to trying to imagine how the Mona Lisa would look if there were none to see. We can only imagine the universe— any universe—as it would appear to our observation, even if it were different from what we actually observe. That there might be other universes, in which different conditions obtain, is a supposition which merits separate discussion, and that it will receive hereafter.

Nevertheless, it has been argued that there might be intelligent beings, different from ourselves, for whose existence other conditions would be needed; so that, had the universe been other than it is, it might still be observed, though not as having the nature we experience. This, if it were true, would make no difference to the self-evidence of a principle analogous to the Anthropic Principle; it would be necessary only to substitute the appropriate form of intelligence for anthropos. Moreover, since the possibility is one which *we* envisage, it could only be so envisaged within a universe in which the Anthropic Principle held. The different conditions necessary for the existence of different intelligences would have to obtain in some region of *this universe*, in principle if not in fact observable by *us*. Further, even if this *per impossibile* were not the case, the assumed intelligences could only observe a universe in which conditions obtained requisite for their own existence. The truth expressed by both the weak and the strong versions of the Anthropic Principle is then surely not as stated by Professor Hawking—"We see the universe the way it is because we exist"[6]— but rather the converse: We exist because the universe is the way we observe it to be, so we could not observe it otherwise.

This restriction on the way we observe the physical world is held to be of importance in science because it exerts what is known as a selection effect. What we observe is conditioned not only by the fact of our existence, but also by the nature and capacities of our perceptive and intellectual faculties. Eddington long ago drew attention to this inevitable involvement of the way we observe and think in what we discover. He wrote: "Let us suppose that an ichthyologist is exploring the life of the ocean. He casts a net into the water and brings up a fishy assortment . . . He arrives at two generalizations: (1) No sea-creature is less than two inches long. (2) All sea-creatures have gills . . . In applying this analogy, the catch stands for the body of knowledge which constitutes physical science and the net for the sensory and intellectual equipment which we use to obtain it . . .

"An observer may object that the first generalization is wrong. 'There are plenty of sea-creatures under two inches long, only your net is not adapted to catch them.' . . ."[7] "At a more mundane level," write Barrow and Tipler, "if a ratcatcher tells you that all rats are more than six inches long because he has never caught any that are shorter, you should check the size of his traps before drawing any far-reaching conclusions about the length of rats."[8]

If this means that what we observe is determined solely by the fact that it is we who observe it, all science would be subverted, for it could then reveal only our own nature and not that of the world at large. Psychologists often tell us that theories (especially philosophical theories) reveal more about their authors than about their subject-matter, and this derogates, and is meant to, from the truth of the theories. If it were universally the case, and were so in principle, all natural science would be rendered nugatory, and only introspective psychology could have any interest of consequence. It is therefore improbable that the scientists who regard the Anthropic Principle as important intend it to have this implication. But if they do not, a great deal of clarification is called for.

There is, moreover, another difficulty about "selective effects" that should warn us of the need for careful distinction between the quest for knowledge and other pursuits. In the case of fishnets or rat-traps, there are other means of investigating the size of the creatures concerned than examining the catch; and in science, where the nature of apparatus or physical limitations on the observer that are in principle surmountable restrict the scope of observation, similar remedies are (at least theoretically) applicable.

Even where observation is confined within irremovable horizons, like those surrounding black holes, or at the visible frontier of a universe expanding faster than the velocity of light, it is still possible to imagine, or (better) to deduce from known facts and laws, what might lie beyond our reach, or how information about it might be obtained. For such selection effects allowance can be made, and they are in some measure remediable.

But in the case of our inherent faculties, their limits are in principle insurmountable, and for them science has no compensation. We cannot by any conceivable means transcend our own perceptual and intellectual capacities, and reference to what lies beyond their scope, as reference to what is in principle unknowable, is meaningless. To speak of a selection effect is to presume a known (or at least ascertainable) range of fact beyond what is selected. But in the case of selection exercised by our sensory and rational capacities, such talk is like Kant's postulation of things-in-themselves, unknowable in the very nature of the case. The presumption now is that our scientific knowledge is irremediably subjective and that the physical reality lies forever beyond our ken.

Any such subjectivism presages epistemological disaster. It would imply that our science, which literally means "knowledge," was not really knowledge at all, because knowledge is claimed on the presumption that it is true of the real world. But, if our scientific observations and lucubrations were an inevitable "selection" resulting from the nature and limitations of our faculties, possibly (but incorrigibly) distorted, they might well be no more than delusions, and solipsism is all too imminent.

Solipsism, however, is so far from being irrefutable (as was alleged by Bertrand Russell) that it is indeed incoherent and self-contradictory, for it asserts the existence of a self *alone*, and thus undistinguished from a not-self and from other selves, whose existence it denies. In so doing, it undermines its own identity, which has meaning only through distinction from an other. In splendid isolation, therefore, no self can exist. Such solitude is impossible even for God, who would be neither infinite nor omnipotent— would not be God—without his creation of the universe. The claim that the Weak Anthropic Principle is scientifically important because it indicates a significant selection effect must be treated with the utmost caution and restraint.

To sum up, as far as we have come, the Anthropic Principle, both in its weak and in its strong forms, can but be self-evident, simply because intelligent life *has* developed within the universe at a definite stage in its history, and to treat observation as exercizing a "selection effect" upon our scientific conclusions is highly suspect.

The Participatory Anthropic Principle is prompted by the Copenhagen interpretation of the quantum theory, initiated by Niels Bohr. According to this interpretation, the uncertainty imposed by Planck's quantum of action upon measurements of conjugate quantities in the observation of quantum systems is not just a contingent feature of our knowledge, but is inherent in nature itself. The Psi-function describing the state of the system defines only a probability amplitude for such quantities, so that the system has no determinate properties until a measurement is made, when the probability amplitude collapses, and only then can the property disclosed be assigned to the system as determinate and actual. Observation and measurement thus become prerequisite to the actuality of all determinate properties of quantum entities, without which they do not exist in reality but are merely potential. The entities involved, however, are the elementary and fundamental constituents of all physical bodies in the universe. It then follows that intelligent beings, through their observation and measurements, must participate in the actualization of the universe at large.

The obvious difficulty with this contention is that human bodies and their sense organs, as well as the measuring instruments and any apparatus they may use for experimentation, are macroscopic objects made up of the microscopic entities that quantum physicists investigate. If the actuality of these entities is conditional upon the performance of experiments, so too must be that of the agents and the means of performing the experiments. Living intelligent beings capable of making observations are, moreover, taken to be the products of biological evolution involving, and continuous with, the physical processes constituted by quantum events. To require that the actuality of these events should depend upon their being observed and measured is to perpetrate a *hysteron proteron*—more plainly, to put the cart before the horse.

It could, of course, be urged that intelligent beings need not necessarily be human. But, unless one postulates the prior creation and continuing preservation of the universe by supernatural beings, the intelligent observers must needs be products of nature and part of the physical world; so that if quantum events, which are primary in every physical phenomenon, all depend on observation for their actualization, the paralogism persists ineradicably for the PAP.

George Greenstein's conception of a symbiotic universe, although intriguing and attractive, is unfortunately incoherent. He avers that the physical world and intelligent life are symbiotic, in the sense that each depends on the other for its existence.[9] But if this were the case, the physical universe could not come to be until observed; and until it existed there could be no observers (because they are consequent upon the generation of life, which is again dependent on the prior occurrence of physico-chemical processes). If physical reality depends on the existence of mind, and mind depends on the prior existence of physical reality, neither can exist unless both can come into being simultaneously. That might be considered a good reason for believing in God, but Professor Greenstein rules out that conclusion as unscientific, and no other theory provides for the creation of intelligent beings simultaneously with the physical products of the Big Bang.

Berkeley's doctrine immediately comes to mind: that material nature is nothing other than ideas (perceptions), apart from which it does not exist at all. The whole of reality consists of observations made by rational beings. This kinship between the PAP and Berkeley's theory has been recognized by Barrow and Tipler. But the Copenhagen interpretation of quantum theory can hardly go quite so far, because it has to account for the measuring instruments as material things. For this reason, some physicists tend to argue that macroscopic objects, being accessible to and determinate for direct perception, are exempt from the quantum indeterminacy.

But here a difficulty arises which does at least open the way to Berkeleyan idealism. For, although measurement requires the coupling of the quantum system to be measured with some macroscopic instrument, once it has been so connected the entire system, including the measuring device, can be regarded as a single quantum system, indeterminate as to its state until the measurement is made and observed. The mathematician, John von Neumann, has shown that not only would this be the case, so that a Psi-function could be written for the whole apparatus, but also, if the apparatus were extended to include a second instrument recording the reading of the first, and then another instrument connected to the second in the same way, the entire chain would become subject to quantum indeterminacy. Where then does it end? A contemporary physicist, Eugene Wigner, has suggested that only when the information enters the mind of the observer does the quantum wave, or probability amplitude, collapse. If so, the theory has become totally and incurably subjectivist, a position of which the dangers have already become apparent.

The Final Anthropic Principle is predicated on the advance of information theory and computer science to the point where von Neumann probes can be developed. These are machines, the theoretical possibility of which has already (it is claimed) been established, that can do everything of which an intelligent human being is capable: they can reproduce themselves, can launch themselves into space, and can travel throughout the universe wherever necessary. Once invented and constructed, they could continue to proliferate and would take over the entire universe. Assuming that the SAP is valid, once intelligent life has emerged and von Neumann probes have been invented and produced, they will continue to keep active intelligence in being, in one form or another, forever.

This claim seems rather speculatively ambitious and smacks of over-confidence. It has been argued that extravagantly sanguine forecasts of the future achievements of artificial intelligence are over-optimistic and that there are insuperable practical obstacles to their attainment.[10] Be that as it may, it is still doubtful whether, as civilization on this planet is proceeding at present, human greed, short-sightedness, and imprudence may not cause the extinction of intelligent beings on earth before developments in computer science have progressed as far as the FAP requires.

Again, the existence of intelligent beings elsewhere in the universe is not to be ruled out, and they may be more circumspect than we have been or are likely to be. So they may survive long enough to invent von Neumann probes and fulfil the conditions of the FAP. However, there is a consensus of opinion among eminent biologists that the existence of extraterrestrial intelligence is exceedingly improbable, because, even on earth, intelligence capable of developing advanced technology has evolved only in one species within two large super-kingdoms of living things, which include tens of thousands of evolutionary lineages; and even among Animalia, with its twenty-five phyla, human intelligence has emerged in only one species among over a hundred, in one order among more than twenty, in over fifty Classes among the chordates. So that even if macromolecular systems have occurred elsewhere in the universe, the probability of their evolving a highly intelligent species is extremely low.[11]

Still more telling is the argument of mathematicians and astronomers, who draw attention to the facts that the universe, and in particular our own galaxy, is old enough for intelligent life to have evolved in a large number of stellar systems and to have reached a stage of development capable of embarking on space exploration like that described above. The motivations (and, for survival, even compulsion) to undertake such exploration are, they maintain, virtually irresistible. In that case, sufficient time has elapsed for them already to have invaded our solar system. That they have not is tantamount to proof that they do not exist.[12]

Consequently, the FAP, on the face of it, looks very dubious, and there seems little, after what has been said above, to recommend any of the formulations of the Anthropic Principle, either scientifically or philosophically. But this is not so, and appears so only because of the way in which the principle has been presented. The scientific facts and theories that have led to its enunciation are of the most profound significance, and they do justify it, though in a rather different way from that advocated by the physicists who have become aware of its validity. They have finally realized the obsolescence of the Copernican outlook with respect to human mentality, and the implications of recognizing the continuity of matter with mind in a unified world. To this aspect of the matter we shall return, but first we must examine some other implications of the Anthropic Principle and alternative ways in which it may be (though misguidedly) advocated—and, likewise, criticized.

MANY-WORLDS THEORIES

A natural reaction to the approach so far made might be that, after all, there is no obvious, cast-iron necessity that the universe should be as we find it. It is surely conceivable that certain key conditions might have been otherwise, or (on some such hypothesis as the Big Bang) initial conditions could have been different. If then the physical environment had been inimical to the emergence of life, there would, true enough, have been no observers; but that it should not have been so inimical and that observers should be present is perhaps simply a matter of chance. Because physical conditions have favoured our existence, we can give an account of the world and of ourselves, and we can raise questions and discuss these issues; but if perchance those conditions had not been favourable, although there would now certainly be nobody to formulate any Anthropic Principle, would not the very possibility of making this supposition validate at least the WAP, even if stronger versions were less defensible? Other possibilities are at least conceivable, and there are possible explanations of the present state of affairs, in which intelligent life has come to be, that do not imply its being inexplicably or inexorably unique. To claim so much we should need to establish that things could not possibly have been different without eliminating any conceivable universe whatsoever. Thus the allegation of a selection effect is not illegitimate, nor is the SAP self-evidently true.

Physicists have put forward models of the universe that support this kind of rejoinder. The universe, they say, might be infinite in space and time. In that case, there could be regions, separated from one another by distances so vast as to prevent inter-communication, in some of which conditions favour the emergence of life, but in others not. In an infinite universe there are bound to be examples of the former, however improbable on thermodynamic grounds. Only in these would there be observers, and they would inevitably observe a surrounding world suited and adapted to their own existence. Hence their presence would produce a selection effect on the results of their investigations.

Alternatively, in infinite time, perpetual changes of physical arrangements would take place, and some that were conducive to the emergence of life would from time to time be bound to arise, even on the assumption of a persistent random shuffling of primary elements. In these epochs, observers could evolve who would experience only conditions amenable to their own existence, under the spell of a similar selection effect.

This theme has been played in several variations. There is evidence that the present expansion of the universe is slowing down. If this deceleration continues and the expansion is reversed to contraction, the universe that began with a Big Bang will end with a Big Crunch. But cosmologists contend that this could be only the prelude to a new beginning with possible new initial conditions. Such alternations might have occurred innumerable times in the past, and could occur again and again in the future, sometimes producing a universe favourable to the emergence of life but at others a different and hostile universe. The implications for the Anthropic Principle are then the same as before.

Yet another "many-worlds" theory has been proposed to counter paradoxes troubling the quantum theory. The Copenhagen interpretation refuses to attribute reality to probable states or properties defined by the Psi-function until experiment has made them determinate. But the many-worlds theory, due to Hugh Everett and John Wheeler, contends that every probable state is actual, each in its own separate universe; so that every quantum event produces a spate of new worlds branching out from it in ever-widening profusion, although only the one revealed by the experiment that measures the specific quantity is observed by the physicist. Of this world, anthropic principles now hold. Our observation is selective and participatory in its creation, and it is the world in which we discover the conditions of our own existence where they must of necessity obtain.

Before considering these variations on the many-worlds theme, a caveat must be issued against a frequent but illegitimate use of the word universe. The Universe is everything that is and ever has or will be; there can be only one. To speak of many Universes is therefore a misuse of the term. If there could be many, they must somehow, and in some sense, be mutually related; otherwise they could not be distinguished, or counted, or regarded as a many. They must therefore constitute a single complex, within which there may be many distinguishable regions or epochs, but these would not strictly be Universes, even if between them no communication of information could pass. If they exist they must have some kind of togetherness. So long as they can be at all conceived and postulated, they will all form part of the all-inclusive Universe. We may then speak of "worlds," as long as we remember that, for all their separability and separateness, they are all parts of the one Universe. That being so, the fact that we inhabit one such world—a fact that cannot (by reason that we state it) be denied, makes any Anthropic Principle applicable to our world automatically applicable to the Universe of which it is inevitably a part.

Accordingly, theories that postulate separate regions in an infinite space, or different epochs in an infinite time, do not contemplate more than one Universe, although they conceive it as infinite. And since they are conceived by human intellects, what was said above about the selection effect is valid against them as against any other theory. We cannot conceive of regions or of epochs that are in principle inconceivable. If we cannot observe them, for whatever reason, we must at least be able to infer to their existence from what we do observe, or else their postulation becomes scientifically otiose. They can be imagined only on the basis of what we already know. If they can be imagined at all, and if they can be legitimately postulated, the Anthropic Principle is relevant to them in the same way as it is relevant to the region that we inhabit and the epoch in which we live.

But serious objections can be brought against the notion of an infinite universe, however subdivided. Infinity is a concept that has given cosmologists trouble ever since Newton, and one which physicists today do all they can to eliminate from their calculations. "That queer quantity 'infinity' is the very mischief," Eddington wrote in 1935, "and no rational physicist should have anything to do with it."[13] Contemporary quantum physicists who find this quantity infecting their equations have invented a method of removing it that they call "renormalization,"[14] and Einstein adopted a similar strategem. Special Relativity establishes an equivalence between matter and energy, and general relativity identifies fields of force with space curvature. Accordingly, matter introduces curvature into space and bends it round into a hypersphere, so Einstein introduced the cosmic constant into his gravitational equation, which eliminates infinity from the resulting model of the universe. The full spatio-temporal extent of the world is now described as finite but unbounded, like the surface of the Euclidean sphere—only, in the case of the universe, the surface has three and not just two dimensions.[15] It is therefore somewhat puzzling to find contemporary cosmologists who adhere to special and general relativity, yet who continue to contemplate a universe infinite in space and time.

Further, these theories often argue for the possibility that order may arise by chance out of random activity in an infinite or very long time. How this is to be reconciled with the second law of thermodynamics is not clearly explained. It must be borne in mind that what is sought is an orderly universe capable of producing life and mind, and, although order can emerge from disorder without violating the second law as long as disorder increases proportionately in the surrounding region, there can be no sur-rounding region if the entire universe is involved. However, it could be argued that the region in question is not the entire universe but only a large though isolated zone. Yet, even if this obstacle could somehow be overcome, there is a more deep-seated difficulty in the very notion of random activity.

In thermodynamics, the random activity presumed is that of molecules dashing hither and thither in a volume of gas or liquid. But molecules are highly structured entities, as are also the atoms of which they are composed. Any random movement must presuppose the existence of some such entities (involving their own order) that can be shuffled around. Prior to such order, there is no discoverable chaos. Present-day particle physics discovers no hard, impenetrable granules. The elementary units are quantum entities that are as much waves as particles, and have been affectionately called "wavicles." They are conceived as wave-packets, superposed waves, at once both energy and matter. Again, waves have structure and are periodic, and prior to them there is nothing except space-time, the metrical field, which itself is an ordered manifold. If it were not ordered it could have no geometry. Where then are we to find the primary bodies that move randomly?

The particle physicist might argue that quantum events are indeterminate, and it is out of these indeterminacies that order has to be built in the macroscopic world. But the indeterminacy affects only the properties of wave-packets and not the system as a whole in which they are embedded.[16] However we look at it, we have to conclude that random activity is always parasitic on some sort of order and cannot have ultimate priority. We shall discover later that it is precisely in the primordial form of order that the conditions for the development of life and mind implicitly reside.

To the Everett-Wheeler many-worlds theory there are even more serious objections. At least *prima facie* it violates wildly Occam's principle of parsimony, *entia non multiplicanda sunt praeter necessitatem*, although some have argued that, by freeing quantum theory from paradoxes of the Schrödinger's cat variety, it reintroduces simplicity. It cannot be denied, however, that it does multiply entities excessively.

More damaging, though, and more relevant to the present issue, is the fact that the parallel "universes" are inaccessible to our observation. They are said to be accessible to observation each in its own world, and it is even alleged that we ourselves are split into the alternative worlds of the quantum potentialities, although we are not aware of this, as we are not aware of the earth's revolution around the sun.[17] This last is surely a false analogy, because we can observe the earth's revolution, if only indirectly, by observing the sun's changing position in the heavens in relation to the constellations of the ecliptic, and its positions at the solstices and the equinoxes. These submissions only provide more stubble to be shaved by Occam's razor, while they fail to exempt the "other worlds" from anthropic considerations, first because they are said to be observable internally, and secondly because they are accessible to us by calculation through the Psi-function.

To the extent, however, that other worlds are actually inaccessible to inspection and devoid of any form of mutual communication, they are simply so much cosmological lumber of no scientific use. At best (if "best" is an appropriate word), they are things-in-themselves, whose existence is postulated without means of verification. They are, in the worst sense, "metaphysical entities." Much less can the theory remove the objections to an anthropic selection effect, or to any of the four versions of the Anthropic Principle.

Suppose there were remote regions of the universe with which we could not communicate, but the presence of which we could infer either from observation of our own region or from our theories. And suppose that in those remote regions the conditions necessary for the development of intelligent life were absent. In them, of course, there would be no observers, and none to raise scientific questions. But how would they differ in principle from those areas of the visible world where no suitable conditions prevail for the development of life or intellect?

There may be millions of galaxies, millions of light years distant from ours, where no life exists and where none can, whatever the cause. They too harbour no observers and provide sustenance to no curious investigators. To us these galaxies appear only as points of light. We can know the details of their nature and composition only by inference. The possibility of more direct communication with them is minimal. Yet they are no more exempt from the implications of the Anthropic Principle than our own galaxy or the outer planets of the solar system. Their existence does not detract from the improbability (such as it may be) of our own existence on earth; rather, if our galaxy is unique in this respect, the discovery of others inimical to life underlines that improbability. On the other hand, in an expanding universe, the existence of remote galaxies may be a condition of our presence here, where millions of years have had to elapse for life to emerge and to evolve intelligent beings. If other worlds exist or have existed of which this is not true, it would make no difference to our own case, nor would our entertainment of the possibility of their existence be exempt from the same selection factor as the rest of our knowledge.

The presumption of many worlds, therefore, does not remove the strictures voiced against the Anthropic Principle in whatever version it is couched; nor does it invalidate it so far as it is legitimate. It may, accordingly, be omitted from further consideration in the ensuing discussion.

WORLD UNIFICATION

How is it that scientists, at this juncture in the development of physics, have come to espouse the Anthropic Principle?

Newtonian physics, as remarked earlier, regarded the material world as external or closed to mind. Observers were seldom mentioned, and the spectacle of the universe was surveyed like a panorama reflected from without into a *camera obscura*. The mind and the mechanism of perception (whatever it might be) were not considered to be the concerns of the physicist, and if they were investigated at all, they were left to the reflections of the psychologist and the philosopher. Mechanics and dynamics were sciences of the material world, the method of which was, indeed, to observe and experiment—to question nature—but only by inspecting its workings, which were to be left unaffected by and aloof from the process of investigation.

The advent of the theory of evolution, we have maintained, built a bridge between matter and life, between environment and organism, by showing that the latter depends on and adapts to external conditions. At first, this had little effect on the outlook of the physicist. It did, however, disturb those who, for religious or other reasons, set the human mind apart, and who, like Bishop Wilberforce, quailed at the thought that men might have descended from apes. Nevertheless, an unbridgeable gap between the physical world and the biotic could now no longer be contemplated, whether or not life depended for its purposive behaviour on the possession of mental faculties.

Quantum theory and Relativity finally undermined the classical dichotomy, because they involved the observer inextricably with what was observed. When it had been established that there was no absolute frame of reference, that every frame was in motion relative to every other, the velocity of the observer affected every observation and every measurement. In E. A. Milne's kinematic relativity theory, the observer was a fundamental factor, affecting every measurement whether of space or of time.[18] And Eddington became so impressed by the involvement of observation with the results it obtained that he went so far as to contemplate the possibility of the mind's being the very substance of the observed.[19]

The effect of quantum theory was even more drastic, for it now became apparent that without the observer the quantity to be measured had no precise magnitude, and the entity observed had no precise properties. But observers are human beings, and human beings are animal organisms, evolved from other species under the influence of environmental pressures. The Anthropic Principle thus enters unavoidably into physical speculation, for the conditions of human evolution are contained in the physical world, the nature of which is known to us only through human observation, and the precise details of which (the Copenhagen interpretation requires) are determined by that observation itself.

Further, the latest advances in particle physics, approaching a unified field theory, have persuaded physicists of the indisseverable wholeness of the physical universe. Human and all other life is included in this whole in ways that make it intimately dependent on the fundamental physical constants of nature. The disclosure of such dependency, more than most other contributing considerations, has prompted contemporary physicists to pronounce the Anthropic Principle. It is this discovery of the unity of the universe, on which physicists now are becoming insistent (although it is not wholly new), that is the really important feature of the principle's recognition. Our next task, therefore, must be to examine the nature of this wholeness and the grounds for the physicists' conviction.

Notes

1. J. D. Barrow and F. J. Tipler, *The Anthropic Cosmological Principle* (Oxford University Press, Oxford and New York, 1986–88), p. 16.

2. Ibid., p. 21.
3. Ibid., p. 22.
4. Ibid., p. 23.
5. *A Brief History of Time from the Big Bang to Black Holes* (Bantam Books, London, New York, Toronto, Sydney, and Auckland 1988), p. 124.
6. Ibid., p. 124.
7. Sir Arthur Eddington, *The Philosophy of Physical Science* (Cambridge University Press, Cambridge, England; Macmillan, New York, 1939), p. 16; cf. *The Nature of the Physical World* (Cambridge University Press, Cambridge; Macmillan, New York, 1929), pp. 239ff. and 264.
8. J. D. Barrow and F. J. Tipler, *Anthropic Cosmological Principle*, p. 2.
9. George Greenstein, *The Symbiotic Universe: Life and Mind in the Cosmos* (William Morrow and Co. Inc., New York, 1988), Ch. 13.
10. Cf. Hubert Dreyfus, *What Computers Can't Do* (Harper and Row, New York, 1972; revised ed. 1979). Roger Penrose raises even more fundamental difficulties with respect to artificial intelligence, questioning the possibility of its ability to account for consciousness and the insight essential to living intelligence: cf. *The Emperor's New Mind: Concerning Computers, Minds, and the Laws of Physics* (Oxford University Press, Oxford, New York, Melbourne, 1989).
11. J. D. Barrow and F. J. Tipler, *Anthropic Cosmological Principle*, pp. 132f.
12. Ibid., Ch. 9, pp. 576 *et seq.*
13. Sir Arthur Eddington, *New Pathways in Science* (Cambridge University Press, Cambridge, 1935), p. 217.
14. Cf. S. W. Hawking, *Brief History of Time*, p. 157, and Michio Kaku and Jennifer Trainer, *Beyond Einstein: The Cosmic Quest for the Theory of the Universe* (Bantam Books, Toronto, New York, London, Sydney, Auckland, 1987), Ch. 4, esp. pp. 67–72.
15. Cf. Albert Einstein, *Relativity, the Special and General Theory* (Methuen, London, 1945; Crown Publishers, New York, 1961), Ch. XXXI, and S. W. Hawking, *Brief History of Time*, p. 136.
16. Cf. Jeffrey Bub, "How to Kill Schrödinger's Cat," *The World View of Contemporary Physics*, ed. Richard Kitchener (State University of New York Press, Albany, NY, 1988); and Errol E. Harris, *The Foundations of Metaphysics in Science* (George Allen and Unwin, London, 1965, reprinted University Press of America, Lanham, MD, 1983), pp. 136–139. The view that uncertainty is an inherent property of physical reality is rejected by Penrose as "certainly wrong"; cf. *Emperor's New Mind*, p. 249.
17. Cf. Paul Davies, *God and the New Physics* (J. M. Dent & Son, Ltd., London, 1983; Penguin Books, Harmondsworth, 1984–86), p. 116–118 and 173–174.
18. Cf. E. A. Milne, *Relativity, Gravitation and World Structure* (Clarendon Press, Oxford, 1935); *Kinematic Relativity* (Clarendon Press, Oxford, 1948); *Proceedings of the Royal Society of Edinburgh*, Sec. A, Vol. LXII (1943–44).
19. Cf. Sir Arthur Eddington, *The Nature of the Physical World*, 239ff., 264, 259: "The physical atom is, like everything else in physics a schedule of pointer readings. The schedule is, we agree, attached to some unknown background. Why not attach it to something of a spiritual nature of which a prominent characteristic is *thought*." Also pp. 264f.: "I have already urged . . . that there is a definitely selective action of the mind; and since physics treats of what is knowable to mind its subject-matter has undergone, and indeed retains evidence of, this process of selection."

2
Wholes

STRUCTURE AND SYSTEM

Before examining in more detail the evidence of wholeness in the actual world, it will be expedient to set out in general the logical principles governing the nature of wholeness and system as such—principles that, as I hope to show, will subsequently be found exemplified in the actual universe, in its physical, its biotic, and its noetic phases. I have made earlier attempts to expound these principles,[1] and here I shall simply summarize the more extended arguments offered in the past in similar contexts, and gather together the conclusions that flow from them.

To say that anything is a whole is to imply that it is not a mere congeries of disconnected and separable items, nor even just a loose collection. It also implies that it is a unity of coherent parts. Every whole is made up of differences that are combined within it to constitute one totality. A purely blank unity is virtually impossible to conceive. A mathematical point having position but no magnitude is not a whole in any strict sense; yet even that, simply to have position, must imply directions in space and distances from other points. Thus, even so abstract a concept is not altogether devoid of internal differentiation. Mathematical space itself is by no means blank, however empty it may be presumed. It is an ordered manifold, involving dimensions, distances between points, possible rotations of lines, distinguishable planes, and the like. Even the simplest of wholes, therefore, is a unity of differences.

Not only so, but the parts distinguishable within the whole, genuinely to belong to it, must be mutually adjusted and must, in some discernible way, intermesh. This is even true of so primitive a whole as a loose collection, because to be a collection at all its elements must have some sort of togetherness involving some sort of mutual effect or influence. A bundle of sticks is a bundle only so long as the sticks are mutually contiguous and each supports those above it and adjusts itself to the shape and weight of the others.

But the word "whole" implies more than this. It implies an interlock between parts that are systematically interrelated (as in a jigsaw), so that their mutual relations are governed by a principle of order or organization that pervades the entire structure. A machine is a good example, or the pattern on a wallpaper, or the design of a work of art. These are wholes of very different kinds and on different levels; they are, however, far from exhaustive. The essential feature is the prevalence of an ordering principle universally determining the interrelations of the elements so that it determines likewise their intrinsic natures, for each must be adapted and adjusted to its neighbours.

Because a whole must be a unity of differences, the elements into which it differentiates itself are finite, each limited and defined by what it excludes and what negates it. We shall see that this mutual exclusion is not (and cannot be) absolute, for reasons already partly apparent; but, being finite, each element is circumscribed by more or less definite boundaries. Nevertheless (or rather, for this very reason), its specific character is determined by what is other than and opposed to it. The elements are opposed because, in some sense, they confront one another. Although opposition is commonly asserted only of the extremes in a series (for example, white to black in the colour scale; hot to cold in a range of temperatures), even consecutive terms in the series are in mutual opposition to the extent that they exclude one another. Yet the opposition is definitive of the contrasted elements, because what anything is in itself depends on that from which it is distinguished, and without such distinction its inner content becomes so vague as to vanish altogether.

A finite entity is therefore in conflict not only with its other, but at the same time with itself, because, while its necessary difference from its other restricts it within its own limits, its very nature is defined by its boundary and by the negating other that it excludes. To assert itself (or to be asserted as itself) within its natural bounds, therefore, it vacillates between denying its other and affirming its inevitable dependence upon it, acknowledging by implication its unity with it. Accordingly, so long as this interdependence and mutual complementarity is ignored or explicitly denied in the assertion of the finite's own identity, its attempt to maintain its independence is frustrated, and it contradicts itself because it disowns that which defines what it claims to be, at once asserting and denying its distinguishing demarcation.

Such self-contradiction is unavoidable, and is ubiquitous among differentiated finites. Yet, while it is endemic, it is not insuperable, because as soon as the mutual implication of the opposites is recognized and their unification in the identity of a more comprehensive whole is admitted, the contradiction is resolved and the opposition reconciled.

Nevertheless, the necessity for differentiation within any genuine whole carries with it the inevitability of internal conflict and provisional disunity, which are the more exacerbated the more the finite elements tend to shun one another and to emphasize their respective exclusiveness in order to maintain their self-identities. This is the consequence of defect in the finite, which is never sufficient within itself, because of its dependence on, and definition by, what it lacks and omits. The result of this conflict is relative chaos and contingency in the unfolding of relations among the elements within a complex, which is overcome, and unity reestablished, only when identity in and through differences is acknowledged, so that the effort of the finite to maintain itself and to persist in its own being succeeds only when it reaches out (so to speak) beyond its own limits to embrace its other and unite with it in mutual complementation in a larger and more comprehensive whole.

Because of their mutual adaptation, interrelated terms within the whole overlap. They all have in common their subjection to the organizing principle, although they must inevitably differ from one another to avoid complete coincidence. Despite this difference, they overlap because of what they share in common, and because of their mutual coordination and interdependence, each for its own specific identity. While they contrast one with another, negate and exclude each other, they are nevertheless defined by their mutual relations and differences, and they are inseparable from one another owing to their mutual implications. This overlap despite difference is what effects their integration into a single whole. For example, the pieces of a jigsaw puzzle are all different, but because their shapes are reciprocally interlocking there is an area of spatial overlap among them, and apart from spatial interlock, the picture or pattern that together they constitute exercises a regulating and coordinating control over their juxtapositions.

Overlap together with integration of opposites in a wider whole involves self-enfoldment, because the wider whole includes the more fragmentary parts, each implying the other in its own self-maintenance. As the implications of the more fragmentary parts are explicated, therefore, what was formerly explicit is repeated along with what subsequently becomes explicit (though formerly implicit), with a consequent complexification of self-enfoldments as the process advances.

To return to the jigsaw puzzle, each convoluted shape implies the contrasting shapes that fit together with it (convex with concave), and when they are conjoined, the conjunction explicates (as does the picture on the surfaces) what was implicit in the separate pieces. So when the fragments are connected, the new combination includes everything that the parts contain as well as what, in each separately, was only implicit but is now made explicit. In a developing organism, such as a growing embryo, the mutual implication of successive stages is more apparent, as is its explicit realization in subsequent phases of development, and the self-enfoldment of the earlier forms and processes to create emerging complexifications is unmistakable: segmentation of primitive cells continues at the stage of specialization and functional differentiation, which again is repeated and internalized in each limb and organ. What ensues is a continuous succession of provisional realizations of the organizing principle (for the embryo, the structure of the mature organism) in a series of wholes increasing in complexity and integration.

Such a series is clearly a system, and every system must be a whole, because the ordering principle is pervasive throughout and integrates all the parts into a single complex. Some wholes, however, are "open" systems, some are self-repetitive, and some are self-representative (like a fractal curve, which repeats itself in every part and on every scale of magnitude). Of this variety and range of systems we shall say more later; what has so far been maintained is that a whole is a concrescent system, united by a governing universal principle of organization that differentiates the unity, expressing (or exemplifying) itself in the mutual adjustment of diverse parts (or elements), which consequently overlap in their reciprocal definition and interdependent relationships.

The overlap of the interrelated elements is double-edged. On the one hand, the identity of each depends on its distinction from the others; on the other hand, its contrast with its correlates is determined by the principle of structure that orders the system and is universal to them all. Adjacent and consecutive elements, therefore, at once exclude each other in mutual opposition, and are complementary each to the other in mutual determination and dependence for their several identities. They are distinct despite overlap, and they are complementary despite contrast and opposition. In each the others are implicit, so each expresses in some degree the universal principle of wholeness; as each requires integration with its neighbours for its own self-maintenance, as it embraces its other in its persistence in its own character (what Spinoza called its *conatus in suo esse perseverandi*), it develops into a more comprehensive whole that expresses the universal principle in a higher degree.

So the elements of the whole are related as opposites, as distincts, as complementaries, and (in successive degrees of adequacy) as exemplifications of the universal principle of order. In this way, progressing by successive steps, from a primitive element up the scale of degrees of more adequate manifestation of the universal principle, the totality that is immanent in every element and every phase of the process develops.

A scale of this kind is dialectical because it proceeds through opposition and distinction, which is at the same time complementarity, interdependence, and mutual identity. In it opposites are united and differents are identified in the unity of the whole they constitute. In the graded scale, each form or phase sublates (supersedes, while it includes, preserves, and transforms) the lower degrees, and foreshadows those yet to emerge. How all this is possible, and how it occurs, will become clearer in what follows, when we examine concrete examples.

It will immediately be obvious that a whole of this nature cannot all be present in any one instant or at any one point. Whatever element or part may be at hand, because of its dependence for its own identity and nature on its relations to other elements, must be unstable within the confines of its own limits. The immanence in it of the organizing principle of the system will impel it to evoke its other and to unite with it to form a more stable entity. Accordingly, the universal principle is necessarily a dynamic principle, forbidding any partial element to rest in isolation from the remainder of the tectonic. It is the *nisus*, the *conatus*, that drives the finite element to transcend its own limits in order to persist in its own being.

The course that it follows traces a scale of forms, the genus of which is the entire complex, explicitly elaborating the principle of organization. Each form is thus a specific exemplification, in its appropriate degree, of that genus. In each the lesser degrees and lower specific forms are all involved, for each is the realization (at least in part) of their potentialities—it is the explication of what, in them, was implicit. The higher degrees are similarly implicit in the lower forms, for the universal is immanent in them all, makes each what it is, and causes it to do whatever it does.

The universal is, accordingly, the explanatory principle throughout the gamut, and each successive form, being the most satisfactory explication of the ultimate totality, up to the point reached on the scale, will provide the key to the intelligibility of all the prior phases in the scale, as well as the most fruitful clue to the nature of what lies ahead. Each form will sublate its predecessors while it presages its successors, until the ultimate totality is reached, which will encompass and explicate, including (while it transcends) the entire scale of forms.

Certain corollaries follow: (1) there can be no ultimately partial whole, or partially systematic complex. Every partial element is at least a provisional whole, because the universal generic principle is immanent in it. Equally, for that reason, the full totality is implied in every partial form, at every stage of development. Because the principle of organization is immanent in each part and constitutes it a part, its very existence as partial must require (as it is itself involved in) the full elaboration of the whole. Without the full circle, if only in presumption, there can be no arc.

(2) Again, for the same reason that any partial element, at any stage, implies the whole, no element, however primitive and lowly, equals zero—there is no zero in the scale.

(3) The series cannot go on forever. It must reach a conclusion—namely,

the totality which is implicit at the beginning and is progressively developing itself throughout. Its principle is one of wholeness; and endless progression never achieves fulfilment, so would never satisfy the immanent universal. Consequently, the scale perpetually tends towards completion and must have an end. In short, neither does the scale include zero nor is it doomed to everlasting insufficiency.

(4) Because the unity of the whole must be a differentiated oneness, the principle of organization must specify itself in finite forms, related, as we have said, dialectically. Internal oppositions and contradictions, the incidents of defect, are thus inevitably involved. Although the oppositions are reconciled and the contradictions resolved as the scale advances, they are ineliminable at finite levels, and are overcome only in subsequent phases of development. Contingency and ineptitude are therefore always one aspect of the progressive self-determination of the whole in its immature stages, with attendant conflict, confusion, error and evil, varying in degree with position in the scale. In the ultimate outcome they will all be wiped out and resolved, but this consummation is not possible, nor properly conceivable, without the process of self-development through which it is achieved, that is, without the phases of its own self-specification (for the very reason that it is a whole only by virtue of its own self-differentiation). The consummation is finally attained only with the unity and coalescence of process and end in the fulfilment of the comprehensive dynamic system.

Such is the character of every whole and of all systematicity. Every whole is a system, however primitive; every system is a whole, structured in accordance with a universal principle of order. That, in consequence, specifies itself in a scale of forms that differ consecutively in the degree of their adequacy to its explicit wholeness. Appearances to the contrary, we shall discover as we proceed, are invariably belied. And the scale of forms is characteristic even of the variety of systems as such. They range from loose and merely haphazard collections of items, through simple arrangements of abstract units, to mathematically structured and dynamically ordered systems. These develop into complexes of more intricately related components and of elements more intimately bound together. The scale continues to the emergence of organic wholes and thence to yet more explicit totalities of more transparently interrelated concepts. Some of these wholes are relatively closed systems; that is, relatively self-contained and self-complete. They are only relatively so, because, apart from the final phase in the scale, none is absolutely whole. Others are open systems; that is, they are configurations or *Gestalten* remaining constant although imposed upon material in perpetual flux. The structure is a relatively closed system although the material basis is fluid. Yet others are self-reproducing and regenerate their own intrinsic pattern either within or outside themselves.

Examples of any or all of these may arise in the course of our investigation of natural and social forms.

UNIVERSALS, ABSTRACT AND CONCRETE

The explicit system of an ordered totality is what Hegel and others have called "the concrete universal," a phrase that conventional formal logicians regard as a contradiction in terms. Any universal, they argue, is necessarily abstract because it is a quality or character common to a number of individual things, which is abstracted from them for the purpose of classification. Its particulars are instantiations of the quality or property thus abstracted, while individuals are made up of a variety of such qualities, and nothing short of the entities in which these qualities inhere is concrete. Most philosophers who adopt this position consider all concrete existences to be particulars, and regard individuals only as logical constructs out of abstract qualities. The qualities themselves are taken to be universals, so on this view all universals are abstract.

The abstract universal represents a class under which its particulars are subsumed, or a set in which they are contained. Classes may contain subclasses, as sets contain subsets. So we can classify things under species, and species under genera, to build up a system of classification. Undoubtedly, this logical schema has a number of useful applications, but its underlying metaphysical assumption is that the real consists of a fortuitous collection of atomic particulars, mutually related only externally. Since the properties and qualities the particular entities possess belong to them only contingently (in this view), even the relation between a thing and its properties is external to the terms, as are relations between distinct things. The particulars can be gathered into sets or classes according to their common properties, but the different classes are related to one another only externally, even when some include others, for the relations make no difference to the terms and are indifferent to them.

This type of metaphysical assumption was encouraged by, and was congenial to, Newtonian physics, which viewed the world as composed of atoms represented as mass-points in external relations (even though, on its own terms, this could not literally be the case). Bodies moved under the influence of forces proportional to the product of their masses and the inverse square of their distance one from another, and were held to be otherwise independent of each other. The relations between them were contingent and did not affect them intrinsically. And, as we have seen, the entire mechanical world was external to the mind, to which it became accessible only through the senses. Empiricist philosophers in the 17th and 18th centuries, especially David Hume, found sensations to be similarly atomic and devoid of necessary connection one with another. So a logic of atomic propositions was well suited to a world of apparently atomic facts. Only particulars were concrete, and universals could only be abstract.

But contemporary physics has abandoned this world-view and has adopted one in which the relations between things and processes and the terms that they relate are intrinsically dependent upon one another, so that they are inseparable in a unified system. Physical entities are thus internally related, that is, related in such a way that the nature of the terms depends on their mutual relations and *vice versa*. Conventional formal logic is therefore no longer appropriate to contemporary physics, nor (as will later appear) to contemporary science in general. What is required is a revised conception of the universal as an organizing principle governing the systematic structure in an ordered whole.[2]

A principle of this kind is universal because its influence prevails through-out the system and is universal to its parts. It integrates them into a single concrete whole, which is generic to the subordinate wholes in the scale of forms into and as which it differentiates itself, and which are therefore its specific forms. The concrete universal subsumes its species and its particu-lars in much the same way as does the abstract universal; but, for the latter, species are mutually exclusive, whereas for the former they overlap, due to its immanence within them. Because it is so immanent in all its particulars, the concrete universal generates its specific phases in the course of its own self-manifestation. It is therefore concrete, and it becomes abstract only when formulated theoretically in abstraction from its actualization in its specific forms and particular manifestations. Similarly, any of its specific forms is abstract, when and so far as the attempt is made to maintain it in isolation from its position in the system and in separation from its proper context or environment.

The concrete universal, then, is a system or whole that generates its own particulars by specifying itself in and as a scale of overlapping forms, which are mutual opposites while they are also complementary distincts as well as degrees of explicit realization of the generic principle of the system. Each of the forms in the scale is a provisional whole. Consequently, its internal structure is similarly dialectical, so that, under analysis, at every level and in whatever part it is scanned, the same dialectical pattern reemerges. Hence every system is a system of systems, and every system is self-representative. This fractal character is not always, in fact not often, apparent on the surface, and has to be discovered by close, sometimes minute scrutiny, just as, in fractal geometry, the self-representative character of the fractal curve involved in natural forms (for instance, the shapes of coastlines or mountain ranges, vascular branching, or the reticulation of fronds in the leaves of a fern)[3] only becomes apparent under detailed analysis.

WHOLES AND RELATIONS

That wholes are complexes of elements in relation must be clear from what has already been maintained. We have already found that the related elements overlap and are mutually determinant. Hence, the relations will be internal to the terms and will determine their intrinsic natures, just as the intrinsic characters of the terms dictate the relations in which they stand.[4] This is so, of course, because the whole is structured by a principle of order that is immanent in all its elements and regulates the relations between them.

But relations of any kind, whether they are in this way internal or are external in the sense that they fall between their terms and leave them unaffected (as is usually assumed in symbolic logic),[5] can be apprehended only by a conscious subject, because what stands in relation must be grasped all together and as a whole. No term alone can comprehend its relation to others, and no subject cognizing them can, in consequence, be confined to any one or any limited group of the terms. It follows that elements in relation must either be objects of some consciousness or must, as a complex totality, be conscious of themselves.

T. H. Green, in the first two books of his *Prolegomena to Ethics*, argues that relations and entities constituted by and dependent for their being on relations can exist only as the objects of some consciousness.[6] This is strictly correct, if by existence one means fully actualized being; but, although a relational complex is only explicit at the level of consciousness, there can be lower levels, prior to consciousness and requisite for its emergence in the order of nature, at which the relations are implicit, so that they exist in potency and latent inter-relativity. These are phases in the dialectical scale prior to mind, through which the natural whole brings itself to consciousness by its inherent *nisus* to self-completion (see succeeding chapters).

Entities that are not conscious can stand in mutual relation only implicitly, and the relations between them, so long as they are not cognized, will be, as it were, in suspense, merely latent or potential. Spatio-temporal relations between points and instants, for example, that are mathematically pure, are, as such, so to speak, moot, apart from their imputation to the spatio-temporal schema by a judging mind. For the terms themselves, they are abeyant and inoperative. If the terms are mass-points or events, the relations between them become causal. Particles affected by gravitational, electromagnetic, or strong and weak forces register their effects not as relations, but simply as attraction or repulsion. If interaction becomes self-sustaining and self-reproductive, it is self-referential and organic, and the terms in relation constitute self-maintaining systems. But while the elements constituting such wholes are responsive one to another, they are not themselves apprised of their mutual relationship. The wholes, as configurations of relations, first begin to emerge in their own right only with sensation, as it contrasts and highlights figure against ground together, in one *ensemble*. They finally come into their own when brought to fruition in perception, to become explicit and manifestly apparent to consciousness.

So regarded, the series of relationships, from mathematical to physical, and thence to organic, sentient, and cognitive, traverse a metaphysical spectrum which, being dialectical, reveals the emergent conscious phase as the actualization of what is only implicit in its progenitors. What is significant here is that the whole, so constituted, insofar as it is a relational complex, is *only* fully actual at the conscious level, because relations, as relations, only exist as cognized. Prior to that, elements that implicitly stand in relation may affect one another or may respond to one another, but only as cognized are they explicitly in relation. As related, they constitute wholes, and (as maintained above) every partial element in a whole, or any immature phase, implies the final completion. Accordingly, what at lower levels is a whole only *in potentia* is real *in actu* only when brought to cognition as a conscious mind.

Since every principle of wholeness is dynamic, impelling its partial elements and rudimentary phases towards completion and fulfilment, any whole must develop so as to attain self-revelation in awareness of itself as a complex of relations. This it does through a process of self-enfoldments producing successive exponential complications, not only mathematical and geometric, but functional and organic. As complexity increases, so integration intensifies, with consequent convergent centreity of concentration as well as a progressive amplification of comprehensiveness, until in consciousness itself the entire universe is potentially embraced.

Thus wholeness, by its very nature, involves dynamic and dialectical self-specification, by way of self-enfoldment (with consequent overlap of specific forms). It tends towards intensification of centreity, increasing self-sufficiency and widening comprehension, and culminates as an all-embracing awareness of an all-encompassing world.

The organizing principle is operative and directive throughout the hierarchy of forms and phases. It brings itself to fruition in self-consciousness, emerging (as will more fully appear at a later stage of our investigation) in intelligent behaviour and interpretive understanding as the activity of thinking. In the process of development, a scale of wholes is generated, each expressing more adequately the presiding principle of order; and what, in its specific degree, each is adequate to, is the final goal, which is explicit awareness of the whole, cognitively set out in explanatory clarity as a comprehensive and coherent experience.

TELEOLOGY

What has thus transpired is that a dialectical process is necessarily teleological, and it is now clear that teleology and wholeness are inseparable
concepts. The word "teleology," however, has habitually been applied in
several different senses, not all of them legitimate, and those which are
unacceptable have brought the notion of teleological explanation into
disrepute.[7] Traditionally, the word has meant "tending to produce an end"
(in the sense of aiming at a goal). This is not objectionable unless it is taken
to imply some sort of causation by future events, which is clearly illegitimate. However, despite the association of final causation with teleology,
which seems clearly to suggest this implication, final causation is not
causation by future events, so much as causation empowered by the
ordering principle of an organized whole.

A teleological process is one in which the whole being generated determines the stages by which it comes to maturity. It is a process directed by
the organizing principle of the whole. Accordingly, a teleological explanation is one which explains the part in terms of the whole, and not *vice versa*.
It is one for which the whole takes precedence, so that the explanatory
principle is that which organizes the system whose parts are the *explananda*.
It is the opposite of reductionism.

To consider teleological process as what tends towards a goal is not
wrong, but it puts the emphasis in the wrong place. The goal is always the
completion or fulfilment of the *nisus* towards a whole. Purposive activity is
only one, and a highly developed, form of teleological activity—one
requiring conscious intention and deliberate choice. This is why it has so
often seemed objectionable to biologists and other scientists to propose
teleological explanations of processes that do not involve consciousness.
Such explanations have (often justifiably) been regarded as anthropo-
morphic. Moreover, the goal of purposive action is not simply its end state.
One does not read (or write) a book merely in order to reach the last page,
but to complete the story. To have the last word in an argument may
sometimes give satisfaction, but the real aim is to make a successful case.
And that is not the last word; it is a coherent and cogent system of evidence
supporting a conclusion.

It is more legitimate to speak of teleological arguments as arguments
from design, for a design is a pattern, and a pattern is or involves a whole;
so the "end" that the argument seeks to prove as the explanation of the
relevant phenomena is the completion, or the achievement, of a "design" or
whole. Thus purposive action, described as action by design, is revealed as
the endeavour to complete a whole and to bring it to fulfilment. Processes
below the level of human purpose, however, may well be teleological

without involving any consciousness, as long as they are determined by the ordering principle of a whole that is generating itself or differentiating itself through them. Arguments from design are objectionable only when they impute conscious intention to beings and processes where no evidence of consciousness can be found, or when they seek to explain the existence or production of nonliving beings (other than obvious artefacts) by the deliberate action of an extraneous consciousness.

We are now in a position to consider a fifth form in which the Anthropic Principle has been formulated: that there exists one, and only one, possible Universe *designed* with the goal of generating and sustaining intelligent observers.[8] This might be abbreviated as TAP, and one can bring objections against it, as against the other formulations; nevertheless, it has far-reaching, important philosophical significance. That there exists only one Universe, of course, should go without saying. The above formulation is evidently phrased so as to counter the implication of a many-worlds theory. Whether other such universes might be possible, however, is now beside the point, because any universe, if it is a whole, must tend to explicate itself in and as an intelligent experience; and if the physical universe, as we observe it, reveals itself as an undisseverable whole, in the light of the foregoing reasoning, its teleological character is ineluctably involved.

To say that "the Universe is designed with the goal of generating intelligent observers" perhaps too strongly suggests that it is the work of an outside agent deliberately aiming at the production of life and personality, human or other. Any such contention could be justified only by further evidence or more extended argument. All that has so far been established is that the design of a systematic whole involves a dynamic principle of order that, by its very nature, tends towards completion of the whole, in and as an intelligent self-awareness. In the end, however, it may well become apparent that the presence in the world of intelligent life at the level of human mentality does indicate a further extension of the dialectical scale and a final consummation in some conceivable necessary self-conscious suprapersonality, no doubt beyond our capacity clearly to envisage, but one that comprehends the ultimate whole, expressing itself in, and specifying the principle of its activity through, the phases of the dialectical scale.

A consummation of this nature would correspond to Anselm's definition of God as "that than which a greater is inconceivable." In terms of a holistic and dialectical conception of the universe, moreover, as I have tried to demonstrate in another context,[9] all the traditional proofs of God's existence can be resurrected, restated, and vindicated as valid.

Our next task is to seek the exemplification in nature of a scale such as has been described in this chapter, and to investigate its course, and its continuation in history and in human intellectual achievement. The principle of systematicity set out above will provide us throughout with the means of explanation, illuminating the relationship between the parts, between energy and matter, between the inorganic and the organic, between body and mind, and vindicating in conclusion all the versions of the Anthropic Principle.

Notes

1. Cf. Errol E. Harris, *The Foundations of Metaphysics in Science* (George Allen and Unwin, London, 1965; reprinted by The University Press of America, Lanham, MD, 1983), Pt. IV, and *Formal, Transcendental, and Dialectical Thinking* (State University of New York Press, Albany, NY, 1987).
2. For a fuller discussion of the discrepancy between formal logic and contemporary science see my *Formal, Transcendental, and Dialectical Thinking*, Chs. 1 and 2. Without repeating what I have written before, I am at a loss to state more clearly the contrasting philosophical implications of classical and contemporary physics. I can scarcely do better than refer the reader to the opening chapters of my earlier book, *The Foundations of Metaphysics in Science*.
3. Cf. James Gleick, *Chaos: Making a New Science* (Viking Press, New York; Penguin Books, Harmondsworth, 1987), pp. 83–118 and 215–40.
4. Cf. Errol E. Harris, *Formal, Transcendental, and Dialectical Thinking*, Part III, Ch. 8.
5. Ibid., Ch. 1.
6. T. H. Green, *Prolegomena to Ethics* (Clarendon Press, Oxford, 1883–1929).
7. Cf. Errol E. Harris, "Teleology and Teleological Explanation," *Journal of Philosophy* LVI, no. 1 (1959); "Science, Metaphysics and Teleology," in *The Personal Universe: Essays in Honor of John Macmurray*, ed. Thomas E. Wren (Humanities Press, Atlantic Highlands, NJ, 1975).
8. J. D. Barrow and F. J. Tipler, *Anthropic Cosmological Principle*, p. 22. I have modified their statement slightly to stress the unity of the universe and the connection of this formulation with the SAP.
9. Errol E. Harris, *Atheism and Theism* (Tulane University Press, New Orleans, LA, 1975), Ch. IV.

3
The Physical Whole

HOLISM IN PHYSICS

In the last decade of the 20th century, the unity and wholeness of the physical universe is hardly a new discovery, although recent advances and newly developing theories in particle physics have aroused more enthusiasm for the idea. At the turn of the century, Planck's discovery of the quantum of action and Einstein's formulation first of Special and then of General Relativity immediately had revolutionary effects. Space and time ceased to be viewed as separable parameters, but were fused together as a single metrical field. Fields of force were conceived as space-time curvature, and an equivalence between energy and matter was recognized. Because matter introduced curvature into space, the metrical field bent round upon itself to form a four-dimensional hypersphere—a finite but unbounded universe. Measurements of length, time, and velocity became interdependent, and a natural unit of measurement was found (by Eddington) to reside in the radius of space-time curvature. Accordingly, an inherent connection was suspected to obtain between this and the fundamental constants of nature, now to be linked inseparably together. Thus the organized structure of the metrical field provided principles of order governing all physical laws and events.

The importance of the field took precedence over that of the particle, and in quantum theory, particle and wave became complementary concepts. The energy system, taken as a whole, thus assumed priority over determination of the exact position, or the precise momentum, of particles within it, so that these properties along with others became conjugate. Pauli's principle of exclusion, prohibiting any two like particles in a system from having the same quantum numbers (i.e., the same state of motion), proved to be a principle of organization, determining the structure of the atom and governing the bonds uniting chemical compounds. In 1959, Heisenberg summed up the new view of the universe that resulted from quantum theory in these words: "The world thus appears as a complicated tissue of events, in which connections of different kinds alternate or overlap or combine and thereby determine the texture of the whole."[1]

31

At the same time, W. D. Sciama proclaimed the unity of the universe (in a book with that name), invoking a variety of evidence that connected centrifugal and Coriolis forces with the motion of the fixed stars, in relation (as he put it) "to some suitably defined average of all the matter in the universe." He similarly related Newton's constant of gravity (G) to the average density of matter in the universe in conjunction with Hubble's constant.[2]

Yet earlier, Eddington had declaimed upon "the wide interrelatedness of things,"[3] and Milne had deduced, from a minimum of accepted data, both Einstein's and Newton's laws of motion, invoking the relation of the moving bodies to the observer and to "the rest of the universe."[4] The internality of relations between physical measurements, physical events, and physical existents became undeniable, and their undissectability from the texture of the all-inclusive, finite but unbounded, space-time hypersphere was widely accepted among cosmologists.

More recently, David Bohm has maintained, by way of discovering a credible interpretation of the quantum theory, that the physical substance of the world is a dynamic totality, which he calls "the holomovement," in which a principle of order is implicated and expresses itself variously in the emergence of phenomena and entities (such as elementary particles), so that, on the analogy of the holograph, the whole is implicit in every part.

Bohm and Hiley have offered an ontological interpretation of quantum theory that is consistent with experimental findings and conforms to Bell's Theorem (see below); it permits the assumption of the reality of the quantum system prior to measurement, so that it avoids both the subjectivism of the Copenhagen interpretation and the extravagance of the many-worlds theory. They conceive the particle as accompanied by what they call a quantum wave field, which satisfies Shrödinger's equation and which, in consequence of its form, affects the movement of the particle, while the particle itself supplies the energy. The analogy offered is that of a radio set, the energy of which is supplied by batteries or mains, while the sounds it produces are generated by the form of the radio wave, whose field is global. The theory is then able to account for the experimental facts, but requires us to regard what is measured and the measuring instrument as a single indivisible complex, within which what is measured comes to be. The results of the experiment are in principle determinate, once the quantum field has been specified, and there is no need to contemplate any "collapse" of the wave function. This theory has not yet been adopted by many physicists, but it illustrates afresh the contemporary trend to interpret the physical facts holistically in terms of the field. In the authors' words: "The fact that each particle responds to information from the entire environment gives a simple and tangible account of . . . the undivided wholeness of the experimental conditions and the experimental results . . . One has to con-sider the whole of the relevant experimental solution, in order to under-stand what happens in each case [in the two-slit, interference-pattern experiment with electrons]."[5]

Further evidence of wholeness issues from current discussions of Bell's Theorem, which, while it seemed to decide the controversy between Einstein and Bohr over the Copenhagen interpretation of the quantum theory in favour of the latter, has now given rise to further grounds for the unification of the physical world. Einstein's doubts about the completeness of quantum descriptions led him, with Podolsky and Rosen, to propound what has become known as the EPR paradox, which for a time encouraged them and others to entertain the notion of hidden variables determining the properties masked by Heisenberg's principle of indeterminacy. Bell's Theorem ruled out the possibility of such covert quantities. Bohm and Aharanov then devised a modification of the EPR experiment which could be performed, and which confirmed quantum theoretical predictions. Now Henry Stapp, by a very cogent argument, based on this elaboration of the EPR paradox, has shown that events separated by a space-like interval are necessarily (even though not causally) bound together by faster-than-light influences.[6] It seems to follow that quantum events are even more firmly embedded in the encompassing energy system than was hitherto suspected.

Events of this kind are described by physicists as "non-local," and there are others that provide further powerful evidence of the interconnectedness of spatially separate quantum systems. Quasi-crystalline structures have been discovered that exhibit hitherto "prohibited" icosahedral symmetry, the assembly of which is necessarily non-local, so that the state of atoms at some distance from the assembly point is involved in the way the interfaces are aligned.[7]

The unifying theories reviewed so far have involved processes that are mathematically describable by linear (and thus soluble) equations. But other, even more extraordinary, testimony to wholeness comes from a new quarter: the investigation of complex dynamic systems (or turbulence), which require for their description non-linear equations. This has given rise to a new department of science embracing mathematics, physics, and numerous other fields—what has become known as the science of Chaos.

CHAOS

The fundamental presupposition of a holistic conception of nature is that whatever one is dealing with is an ordered system. Nothing less can be a genuine whole. As previously maintained, neither a blank uniformity nor a dimensionless point is a whole, though it may serve, for some purposes, as a unit. Now, there are two objections, or lurking misgivings, against holism that have been raised from time to time. One is that the basic physical reality could be some form of purely random activity, out of which order might arise by chance simply through the operation of the law of averages, or of large numbers; and at least in its early days the quantum theory seemed to suggest this. The second is that reality might be sheer chaos, and that the appearance of order is an imposition upon our disorderly sentient experience by an inveterate tendency of the mind to organize its data, as the condition of understanding, coupled with the gross limitations of our senses. I have always thought, and have argued above, that both of these objections can be conclusively refuted on logical and philosophical grounds, and that randomness is always parasitic on order. Now, almost by accident and mere serendipity, new mathematical revelations have demonstrated in quite unexpected fashion that this is the case, that chaos is simply a superficial mask of the most intricate and entrancing forms of order and pattern, and that its occurrence in nature is determined mathematically. These revelations have been made in the course of new developments in the study of complex dynamic systems.

The development began in the 1960s with the attempt by a mathematically minded meteorologist, Edward Lorenz, to map, with the help of a primitive computer, a simulated weather system governed by twelve non-linear equations. The omission of the last three of six decimal places in one of the parameters, a change of only one part in a thousand in the value of the input, produced an unexpected and extreme aberration in his output graph, revealing what is called sensitive dependence on initial conditions, or, more jocularly, "the butterfly effect." The suggestion is that the beat of a butterfly's wings in China would subsequently alter storm systems in the Caribbean. This assumption of a global interdependence of events puts one in mind of the habit of Professor Brand Blanshard of Yale, years ago, who, when beginning his introductory lectures in philosophy, would throw a piece of chalk at the blackboard and turn to the class saying: "I have just altered the coast-line of China." The sensitivity to initial conditions thus revealed underlines further "the wide interdependence of things" already apparent from other considerations.

Lorenz went on to investigate the unheralded introduction of turbulence in his system by examining the phenomenon of convection in a fluid medium, using a very simplified non-linear, three-equation model. He traced this as a three-dimensional graph, which he found moved in a double loop that never repeated itself exactly, but was restrained within a distinctive pattern (oddly enough, shaped like a butterfly). Such patterns later came to be attributed to "strange attractors," to which I shall return in a moment. Thus he began a trend away from looking locally for mechanisms and towards seeking pattern amid randomness.

Further investigation of fluid turbulence and complex dynamic systems, with the help of computers, led Philip Marcus successfully to simulate a turbulent system that reproduced a steady shape similar to the great red spot on Jupiter, which, it transpires, is an instance of stability within chaos. Marcus had found that a complex system could give rise to turbulence and coherence at the same time.

James Yorke, examining the results of running a simple non-linear equation for population growth, found that as the key parameter increased, the curve for steady growth bifurcated with one for alternate growth and decline, and then each branch bifurcated again, until a point was reached at which the sequence seemed to degenerate into chaos. However, whenever and wherever the chaotic area was enlarged, it revealed the same (or virtually the same) pattern of bifurcations on a smaller scale. This was, in effect, a discovery in nature of a self-representative system, as conceived in the last century by Dedekind and Josiah Royce. It was, in fact, a fractal curve, such as Benoit Mandelbrot was to study with results of incredible fruitfulness—a curve in which any portion, on any scale, reveals the same pattern.

Mandelbrot discovered this characteristic in a vast array of differing contexts: market price fluctuations, physiological and morphogenetic processes, and anatomical structures in animals and in plant forms, and he proclaimed that fractal geometry was the universal geometry of nature. The important point is that what appears as a jagged and broken shape turns out to be a fractal curve embodying a hidden organizing structure. It is similar to what David Bohm has maintained is characteristic of the physical world as a whole, in which an "implicate order" is enfolded in the "holomovement" of physical activity, such that an organizing principle is implicit in every point in space-time, as in a hologram—the whole immanent in each part.

Problems of phase transitions (e.g., between solid and liquid, liquid and gaseous states, the magnetizing of metals, and the like) proved tractable when treated in terms of fractals, revealing that they all obeyed the same rules. Mitchell Feigenbaum discovered a constant number involved in the periodic repetition of similar patterns on different scales, which indicated a universal law governing "strange attractors," which turn out to be, or at least to involve, fractal curves.

A strange attractor is a dynamic structure towards which all motions in a complex dynamic system tend at specific degrees of energy flow. Not only Lorenz's butterfly pattern (commonly called "the Lorenz attractor"), but numerous others, structured probably by several such attractors, fall into this category. The term indicates an assumed or apparent centre towards which the motions within the system are drawn.

When the equations representing the system are plotted on what mathematicians call the complex plane (the coordinates of which represent "real" and "imaginary" numbers), they trace patterns, which can be represented in computer graphics and which display the most varied and entrancingly decorative shapes. They all turn out to be fractal forms, and Mandelbrot discovered that, from a very simple formula, he could generate one image that proves to be the source of all the rest and serves as a guide to their emergence in complex dynamic systems. This is known as the Mandelbrot set. It is a fractal set of extreme complication that yields, when displayed in (arbitrarily) coloured computer graphics, fantastically beautiful and intricately diverse patterns, as viewed on different scales.

Thresholds governing phase transitions are the boundaries between the pulls of strange attractors, and these boundaries are themselves fractal forms, which (as it turns out) correspond to the shapes and dispositions of natural objects: fern leaves, snowflakes, jellyfish, the shells of foraminifera, and so on. Just as fractals serve to explain the ramification of blood vessels in the body and alveoli in the lungs, the branching of plant forms, and dozens of other natural structures, so this new branch of science (a hybrid of many disciplines, involving a form of experimental mathematics) has found application to diverse physiological and pathological processes, like circadian rhythms and cardiac fibrillation.

Strange attractors have been seen as conflating order and disorder, so that what we take to be random and disorderly is the effect of a principle of order, which reemerges from the turbulence that results from its operation. In terms of communication science, they have been interpreted as generators of information—what biologists have been seeking for decades, the source of decreasing entropy. But most important of all is what scientists pursuing these new developments have repeatedly stressed: that they seek the whole, the overall structure, apart from which it is futile to examine and try to understand the parts. "For them," writes James Gleick, "chaos was the end of the reductionist program in science."[8]

THE THEORY OF THE UNIVERSE

There can be no question that the conception of the physical world entertained by contemporary physicists is thoroughly holistic. What has become well-nigh a general consensus is eloquently expressed by Fritjof Capra, who has persuasively argued that there is a close analogy between the emergent conception of the world in the physics of the late 20th century and that of the ancient mystical thinkers of the East. He writes:

The basic oneness of the universe is . . . one of the most important revelations of modern physics. It becomes apparent at the atomic level and manifests itself more and more as one penetrates deeper into matter, down to the realm of subatomic particles . . . As we study the various models of subatomic physics we shall see that they express again and again, in different ways, the same insight—that the constituents of matter and the basic phenomena involving them are all interconnected, interrelated and interdependent; that they cannot be understood as isolated entities, but only as integrated parts of the whole.[9]

Let us consider some of the facts and theories that contribute to this opinion, and ask what sort of a whole they reveal.

To the end of his life Einstein had sought a unified field theory, an aim he had conceived, along with Schrödinger and Weyl. What seemed insuperable difficulties in uniting relativity with quanta in a quantum field theory for some time brought the idea into disfavour among physicists, because of its seeming unfruitfulness. But when the evolutionary (or Big Bang) theory of the universe prevailed over the steady-state theory, and quantum physicists began to speculate about how elementary particles would behave at enormously high temperatures (like those immediately after the Big Bang), the unified field theory came back into view. Physicists now had to deal with velocities approaching the speed of light, and so relativity had to be combined with quantum theory in order to solve the problems that arose. It was this unification of relativity and quantum theories that Einstein had sought, and the pursuit of a unified field was therefore resumed.

The concepts characterizing the new theories are symmetry, gauge symmetry, and supersymmetry, gravity and supergravity, strings and superstrings. To expound them accurately one must be a physicist and mathematician; to explain them intelligibly to the layman, one must be a consummate and ingenious stylist. I can claim neither the qualifications nor the necessary talent, and all I can hope to do is to give a brief and summary account of their content. The implication in these theories, however, of a unified physical world is quite unmistakable.

It was said above that, in modern physics, field had taken precedence over particle; but this is only one side of the picture. True, quantum theory has united particles and waves, but for that very reason, and because even radiation is quantized, at the micro-level field and particle become indistinguishable. Because forces are transmitted in small packets, there are characteristic particles for each field. That of the electromagnetic field is the photon, that of the gravitational is the graviton. And there are two other forces: that which holds the atomic nucleus together (the strong force) and that which operates in radioactive decay (the weak force). Each of these has its own peculiar particle.

Particle physicists envisage interaction between particles (when they collide, or when particles of different electric charge attract one another) as the transmission between them of so-called exchange particles. These are known as virtual particles because, except in special circumstances, when they become detectable as waves, they cannot be directly measured, although their effects can. As the result of collisions between actual particles at high energy levels, virtual particles may become actual, and then their tracks can be seen or photographed in such devices as a Wilson cloud chamber.

Further, in quantum theory, energy and time suffer the same sort of indeterminacy relation as do position and velocity, so that the energy of a system considered for a very short period will fluctuate incalculably. Accordingly, at the micro-level, energy may appear, as it were, spontaneously, in "empty" space, due to surges powered by the Heisenberg uncertainty. These energy surges, in observance of Einstein's equation,

$$E = mc^2$$

can produce virtual particles in fleeting succession that last only the minutest fraction of a second. In this way, space becomes a sort of shifting miasma of ephemeral virtual particles through which the subatomic particles of matter constantly move. This is what constitutes their fields. Once again, space-time and energy are seen to be inseparable aspects of a single reality, as are energy and matter, wave and particle.

Electrons orbiting round the nucleus of an atom are held in their orbits by an electrodynamic force which obeys the inverse square law. But the nucleons (protons and neutrons) are held together by much stronger forces, which must be able to overcome the mutual repulsion of two positive electric charges. Here the inverse square law can no longer be applied, if only because the particles are in such close proximity that the values calculated would approach infinity. This is what wrecked the first attempt to combine quantum and relativity theories, in what was known as quantum field theory—its equations spawned infinities that scientists did not know how to eliminate.

When particles collide they tend to scatter, and the numbers that contain the information about such scattering constitute what is called the S-matrix. But when the particles are moving at very high velocities, so that special relativity has to be combined with quantum theory to calculate their motions, it was found that the S-matrix was plagued by infinities. Richard Feynman, however, succeeded in 1949 in modifying the S-matrix with the help of gauge symmetry so that the infinities cancelled one another out. This is known as "renormalization" and it enabled Feynman, by exploiting gauge symmetries, to unite the photon and the electron in a quantum electrodynamic theory (QED). It achieved no union, however, between electromagnetic and weak or strong forces, much less between any of these and gravity.

Gauge symmetry has been explained by Paul Davies on the analogy of two climbers on a cliff face, one of whom ascends vertically and the other by traversing along a zigzag path. The distances are very different, but the energy expended in either case is the same.[10] Other symmetries have been revealed among elementary particles. Just as the left hand is symmetrical with the right, or the human body with its mirror image, so the mathematical representations of certain different particles (for example, the proton and the neutron) are found to be symmetrical. This has enabled physicists to introduce order into what, at first, seemed to be hopeless confusion.

The discovery in quick succession of pions and muons, in two varieties, and then of dozens of new subatomic particles, bemused researchers and theorists who had been used to the simple regime of electron, proton, and neutron, with the tolerable admission of the positron. But the recognition of symmetries enabled them to classify and relate the growing throng of particles as mesons (including pions), leptons (those similar to the electron but not all charged), and hadrons (the heavier particles, such as protons and neutrons, which decay into leptons, and are known as baryons). Leptons are mostly subject to the weak force but, if charged, are subject also to the strong, while hadrons are strongly interacting.

The use of the mathematical theory of groups led researchers to conceive of a higher symmetry underlying those already utilized, and guided them to the discovery of quarks, which came to be seen as the genuine elementary particles, although they disclosed a number of varieties (appearing in six "flavours" and three "colours") and were not isolable. They could be correlated with the leptons, but were subject to the strong force, the messenger (virtual) particles uniting them being called gluons. All were found to be connected by symmetries.

Attempts to unify the weak force with the electromagnetic, via a presumed exchange particle known as the W-particle, failed at first because the proposed equations were not renormalizable. But the use of a new and more sophisticated gauge symmetry (the Yang-Mills theory, which eventually proved to be renormalizable) and the assimilation of the W-particle to the photon brought success, and the electro-weak force could be accepted as one.

An extension of the unified gauge theory next afforded the means to the unification of the electro-weak force and the strong force. This has become known as the Grand Unified Theory (GUT), which leaves only gravity to be included to give final unification. But gravity was still pestered by infinities and presented further daunting obstacles. Theories have been entertained involving what is termed super-gravity, but the one that is emerging as apparently the most promising is the superstring theory introducing supersymmetry.

The string theory was first introduced during the 1960s in an attempt to bring order into theories of the strong interaction and the proliferation of hadrons. It was based on very stringent conditions set down by Geoffrey Chew for determination of the S-matrix, and, because of his insistence on strict consistency, it became known as the "bootstrap theory" (holding itself up by pulling on its own bootstraps). This S-matrix theory was independent of the different types of elementary particle, treating them all on a par. It dispensed with Feynman diagrams and did not require renormalization. The introduction of quarks and gluons, however, distracted the attention of physicists from the idea, until the mounting difficulties of reconciling relativity with quantum mechanics in very high energy situations brought it back into prominence.

It was next suggested that the elementary units of matter are not point particles but are extremely minute strings, and that the different kinds of particle are merely different vibrations or resonances on a string, like the different tones and overtones on a violin (or piano) string. This hypothesis has much in common with S-matrix theory, because it regards no form of "particle" as more fundamental than any other; yet, like quantum field theory, it still contemplates elementary units of matter.

As particles can collide and diverge, so strings can combine and form a single string, or separate again to reform divergent strings. Some strings are regarded as "open," that is, they are open-ended, the lowest vibration on such a string corresponding to the photon. Others curl round into a loop and are closed strings. The lowest vibration of the closed string corresponds to the graviton. The theory is able to accommodate every known type of particle, including those corresponding to the four kinds of force, so that it promises to be (and is projected as) "the theory of the universe."

Strings are one-dimensional entities, but projected in time they have what physicists call world-lines and become two-dimensional "sheets." In the case of closed strings, these world-lines are tubular, three-dimensional pipes, which can merge and separate like veins in a leaf or blood vessels in a living body. The exchange of a virtual particle between two others (say, electrons) may be represented schematically in an H-shaped tubular structure.

The theory had some drawbacks. It was at one stage found to be consistent only in twenty-six dimensions, which seemed excessive. But then a modified and more sophisticated form proved consistent in only ten dimensions. Of these, it has been suggested, all but four are rolled up in a minute range smaller than the size of a proton, while the rest are spread out more or less "flat," and are those we experience. The explanation given for this is that in its earliest moments the universe was, as a whole, of minute size and in ten dimensions. It also existed in a state of extremely high energy, with temperatures of billions upon billions of degrees. In these circumstances, all four forces were reduced to one. But in such a state the universe was highly unstable and exploded in the colossal blast that first separated a four-dimensional space-time from the other six dimensions, and then started an inflationary motion which, as it were, ironed out the kinks in four-dimensional space-time. The expansion and consequent fall in temperature then produced phase breaks, which separated out the four forces now observable and, in due course, the range of "particles" that have been discovered.

Beginning with a single, miniscule, ten-dimensional coil, the world divides into two parts, one of which is four-dimensional; then space-time expands with extreme rapidity. As E. A. Milne asserted roughly half a century ago, "the spreading light wave extends space." Thus, the physical universe can be represented as a single string stretched to its outermost limits, on which the multifarious forces and particles are different harmonic vibrations, mathematically related. As David Bohm contends, it is a "holomovement" enfolding an implicate order, which manifests itself in the diverse physical entities and forces that we observe.

The superstring theory involves a highly advanced and sophisticated mathematical algorithm based on group theory, and has been found to include every known form of symmetry, eliminating all infinities and anomalies. Nature, it has been maintained, not only tolerates but demands symmetry, so that to remove all divergences and anomalies *only one* theory is admissible. It is no longer possible for theorists to contend that they can write down any number of consistent theories, any of which would conform to relativistic principles and to quantum mechanics. They now find that only one will serve, if anomalies and infinites are not to be encountered.[11] Ultimate coherence seems finally to have triumphed as the indispensable condition of a satisfactory explanation of the physical world.

It is now some thirty years since Heisenberg predicted that physicists would discover "a fundamental law of motion for matter, from which all elementary particles and their properties can be derived mathematically."[12] Today, this prophecy seems about to be fulfilled. The physical universe is proving to be a seamless texture of inseparable events and entities, organized in accordance with a universal principle that specifies itself in innumerable forms, all of which can be deduced from it, once it has been discovered.

THE PHYSICAL SCALE

Coherence, elegance, and symmetry, the criteria of beauty and truth sought by the mathematician and the theoretical physicist, seem now to be within the reach even of the experimentalist, as plans materialize to build a super-cyclotron, the super-conducting super-collider (SSC). The biggest machine ever to be built in pursuit of pure science is intended to generate temperatures and energy levels approaching those of the early universe, so as to provide evidence either to confirm or to dispose of current theories. Meanwhile, the philosophical significance of the already apparent structure of the physical universe may be explored—the task of the present chapter.

The world picture implied in the theories outlined above is one of a single unbroken whole, governed by a principle of organization universal to a self-generating system, in which this universal specifies itself in a scale, a series of forces and entities, ranging from the simplest to the most complex and opening the way to a further development on a higher level of organic wholeness.

The universal principle itself has, so far, only been adumbrated, and cannot as yet be precisely formulated; but if and when it is, it would still remain no more than an abstract formula, apart from its active self-specification in, and as, the physical world. It is in its actualization and in the outcome of its self-development that its true nature would finally be revealed. What this is, therefore, must be reserved for later consideration. First let us trace the stages or degrees of realization discovered in the physical universe.

We begin with the multi-dimensional metrical field, wrapped up into a minute compass, but enfolding the implicit order of the whole. This disrupts into an inflating four-dimensional whole and a contrasting six-dimensional complement. They are in contrast, yet are complementary as the basal structure of the string complex yet to emerge, and are both instantiations of the universal principle of order.

As the field expands, it differentiates further into four distinct forms of energy, forces that manifest themselves both as waves and particles (or wave-packets). Energy and matter are contrasting opposites, yet mutually inseparable complements, which form new quantized phenomena, ranged in a scale of particles: hadrons opposed to leptons, fermions to bosons, charged particles to uncharged, positive to negative. They are still mutually complementary, forming specific wholes, such as protons and neutrons, constituted of diverse quarks.

These entities display various symmetries that are, in effect, identities of opposites, and they combine in overlapping specific forms to create larger wholes. Exchange particles contrast with matter particles, yet they overlap as species and appear to combine and separate under varying conditions. The overlap is a sort of self-enfoldment. Energy may be represented as space curvature; particles (or wave-packets) are formed by superposition of waves; mesons and baryons are made up of overlapping quarks; nuclei are constituted by the overlap of protons and neutrons; atoms by that of electrons and nuclei; molecules by that of orbiting electrons in the combining atoms. The self-enfoldment of successive manifestations of the underlying matrix produces a range of specific forms that consequently overlap and complement one another, while they are mutually in contrast and opposition, so that they constitute a scale of degrees of wholeness, or integration, which specifies the universal principle of order progressively more fully.

Symmetries discovered by Murray Gell-Mann and Yuval Nee'man enabled them to arrange hadrons in an eight-fold scale not unlike the Mendlejeff table of chemical elements. Thus there emerge scales within scales, systems within systems, as the dynamic principle of order, immanent throughout, differentiates itself in the micro-world. The scale of physical forms begins with energy expanding to create space-time and differentiating itself into specific forces correlated with elementary particles. These are deployed in a scale of increasing complexity, opposition, and complementarity, from quarks to hadrons, which then, in contrast with, and complementary to, electrons, combine to form atoms, and they proceed to embark upon a further scale of complexification. The general principles laid out above, in Chapter 2, are maintained and exemplified throughout the physical universe.

Atoms, as they become more complex, gather more nucleons and orbiting electrons to form heavier substances, and these are ranged in a new scale of chemical elements, as set out in the Mendlejeff table. Atoms join together to form molecules, which are again structured in accordance with the laws of electromagnetic and strong interaction. The elements combine variously, to produce compounds, their chemical bonds determined in all cases by the organizing principle embedded in the atoms themselves, Pauli's Principle of Exclusion. Under appropriate conditions, molecules coalesce to form crystals, within the so-called leptocosm of which they are arranged in accordance with the same laws and principles, exemplifying the original universal at a higher level. More complex molecules combine into polymers, and the scale proceeds to organic substances, such as proteins and nucleic acids, necessary for the emergence of life.

So there is a continuous scale of "complexification" from space-time to those forms transitional between the inorganic and the organic. It is a dialectical scale of opposing, yet overlapping, specific forms, which differentially instantiate a single universal principle of order in continuously increasing degrees of complexity and integral wholeness.

But this is only half the picture, which is paralleled by the other half—the macrocosm of the expanding universe.

As the universe expands, and with the creation of ions and atoms, gas clouds form and, under the influence of gravity, contract, split up, and form stars and other bodies that are grouped together in galaxies. In accordance with their masses, stars evolve in diverse ways. Some, known as main sequence stars, have longer life histories; others, less stable, are shorter lived, and their transitions are more violent. Main sequence stars burn hydrogen to helium in internal atomic furnaces, radiating light and energy. Some collect around them planets and other satellites. When their atomic fuel is burned up, they collapse under the force of their own gravity. In some cases, they may explode, as supernovae, to form neutron stars, or finally, if their masses are critical, black holes. Only stars of mass similar to our sun belong to the main series. Those of greater or lesser mass have different histories.

This macrocosm of stars and galaxies stands in contrast to the microcosm of subatomic particles, yet, as we shall see in more detail presently, the two scales are inseparably linked and are indispensably complementary each to the other. Not simply are atoms and their nuclear forces essential to, and participatory in, the creation and the life-history of stars, but only as the result of the processes of evolution in main sequence stars can the heavier elements ever form, and only as a result of the explosions of supernovae are they spread abroad in space. If there were no main sequence stars, there could be no planets like the earth, with atmospheres containing oxygen and carbon dioxide, or lithospheres containing the heavier metals. It follows that there could then be no life, no biosphere, and no observers. In short, the microcosmic sequence from hydrogen atoms to macromolecules depends intimately upon the macrocosmic sequence of stellar evolution.[13]

Once again, we find a scale of forms: planets, stars, galaxies, galactic clusters, continuous right up to the final hypersphere. The scale is continuous because, according to general relativity, the presence of matter introduces curvature into space, bending it round to create the hypersphere. Moreover, as noted earlier, the space-time continuum itself is created by the pervasive activity of energy and its complementary matter waves. Thus, macrocosm and microcosm are dialectically opposed yet equally interdependent and complementary, together forming one systematically integrated totality.

The physical world, however, is only the base on which the biosphere rests, as the lifelike statue of Dionysus stands upon its marble pedestal. Or, perhaps a better simile would be that of an ocean in which living creatures swim. Hegel maintained that the earth was an organic whole, in which, at a certain stage, life broke forth in scintillations, like the stars of the heavens.[14] Taken as a metaphorical representation of the relation between the physical and the biotic spheres, this is not an inappropriate conceit. Our next step will be to take a closer look at the physical base to see how intrinsic and indispensable are its essential characteristics to the existence and support of living beings.

Notes

1. Werner Heisenberg, *Physics and Philosophy* (Faber, London, 1959), p. 96.
2. W. D. Sciama, *The Unity of the Universe* (Doubleday, Garden City, NY, 1959), esp. Chs. VII–IX.
3. Sir Arthur Eddington, *The Expanding Universe* (Cambridge University Press, Cambridge, 1933), p. 120.
4. E. A. Milne, *Proc. of the Royal Society of Edinburgh*, Sec A, LXII (1934–44).
5. Cf. David Bohm, *Wholeness and the Implicate Order* (Routledge and Kegan Paul, London, 1980); and D. Bohm and B. J. Hiley, "Non-relativistic Particle Systems," *Physics Reports*, 144, no. 6 (1987).
6. Cf. Henry Stapp, "Are Faster-than-light Influences Necessary," *Quantum Mechanics versus Local Realism—The Einstein, Podolsky and Rosen Paradox*, ed. F. Salleri (Plenum Press, NY, 1987); "Quantum Theory and the Physicist's Conception of Nature: Philosophical Implications of Bell's Theorem," *The World View of Contemporary Physics: Does It Need a New Metaphysics?*, ed. Richard Kitchener (SUNY Press, Albany, NY, 1988).
7. Cf. Roger Penrose, *Emperor's New Mind*, p. 436.
8. James Gleick, *Chaos* (Viking Press, New York; Penguin Books, London, 1987), p. 304. In this connection, of particular interest is the suggestion made by Roger Penrose that the paradoxes of quantum theory indicate its provisional character, demanding a more complete theory that would involve non-linear equations to which the linear superpositions of quantum mechanics would prove to be approximations (as Newton's linear equations of gravity proved to be approximations to Einstein's non-linear relativistic equations).
9. Fritjof Capra, *The Tao of Physics* (Fontana, London, 1975), p. 142.
10. Cf. Paul Davies, *God and the New Physics* (Dent, London, 1983; Penguin Books, Harmonsworth, 1984), pp. 156–157.
11. Cf. Michio Kaku and Jennifer Trainer, *Beyond Einstein: The Cosmic Quest for the Theory of the Universe* (Bantam Books, Toronto, New York, London, Sydney, Auckland, 1987), pp. 108f. J. D. Barrow and F. J. Tipler, *Anthropic Cosmological Principle*, pp. 257–258, citing S. Weinberg, *1979 Nobel Prize Lecture* (Nobel Foundation, Stockholm, 1980), J. A. Wheeler, *Monist*, 47 (1962), p. 40, and B. S. De Witt, *Phys. Rev.*, 160 (1967), p. 1113.
12. Werner Heisenberg, *Philosophical Problems of Nuclear Science* (Faber, London, 1952), p. 103, and *Physics and Philosophy* (Faber, London, 1959), p. 60.
13. Cf. J. D. Barrow and F. J. Tipler, *Anthropic Cosmological Principle*, p. 246.
14. Cf. G. W. F. Hegel, *Naturphilosophie, Enzyklopädie der philosophischen Wissenschaften*, Sect. 341.

4
Physical Prerequisites of Life

INDISPENSABILITIES

Intelligent life as we know it can exist only within a relatively narrow range of temperatures. The universe, according to current theory, began with temperatures of millions of degrees and, in the course of its expansion, has cooled off to levels far below that at which life can be engendered. Life, therefore, can exist only in regions warmed by some source of energy, like the sun, a hydrogen-burning star of the main sequence variety. Such a star could have evolved, from contracting clouds of cosmic gas, only after a long period of time, little less than the age of the universe itself. Our own sun, it is estimated, has been burning for about 4.5×10^9 years. The time in which intelligent life could evolve to the stage we have reached is not much less (about 5×10^8 years). In a time lapse such as is necessary for these developments to have occurred, the universe would have expanded to the extent that we now observe. So a physical world of at least 1.8×10^{10} years in age and as many light years in radius is prerequisite to the emergence of intelligent living beings capable of observation and reflection, thinkers able to ask questions about themselves and their environment, and capable of so organizing their ideas as to find answers to them.

Obviously, these facts are all necessarily connected, and it is no cause for astonishment that physicists discover these quantities to be what is required if intelligent life is to inhabit a planet like the earth. We are aware that we do exist here and now, and are apprised of this fact by our very awareness. This being so, it could not be otherwise than that the conditions necessary for our existence should obtain and that, accordingly, we should, on investigation, find that they do. That is not of great significance philosophically or otherwise.

The universe is as it is, not *because* we exist and observe it to be so. Unless our presumed knowledge and much vaunted science is utter illusion, we observe it to be so because *it is* so (within reasonable limits of error); and because it is so we can exist and are able to observe it. What is significant, however, is that physicists discover the principles determining the structure of the universe to be so finely tuned, and the relations between its parts so minutely adjusted to one another that the emergence of intelligent life is incompatible with any other possible arrangement of things and events. Were we to find that the universe could not have been other than it is, and that its being so is inseparably bound up with the emergence and evolution of life forms, that would be of the most profound importance.

INCOMPATIBILITIES

The currently accepted theory of the universe is that it began some eighteen thousand million years ago with a vast explosion. Its present age, size, and rate of expansion all depend upon the relation between the forces of gravity and of the initial propulsive outburst. Had the latter been weaker (Paul Davies tells us), the cosmos would rapidly have fallen back upon itself and contracted to a point. Had it been stronger, the cosmic matter would have dispersed at such speed that galaxies could not have formed. A difference of one part in 10^{60} would have been sufficient to bring about either of these two results. The precision of the actual balance of forces, therefore, necessary for the presently observed universe to exist matches that required to aim a rifle bullet at a one-inch target twenty billion light years away.[1] In short, the present structure of the universe is perhaps the ultimate example of sensitivity to initial conditions.

The minutest difference in either of these forces would have resulted in the self-annihilation of the universe; for if it had re-collapsed upon itself it would have become an immense black hole, and Stephen Hawking maintains that in black holes matter probably eventually evaporates away altogether. If the elementary particles had dispersed so rapidly that no stars or galaxies could form, the universe would consist of a sort of tenuous gas of electrons and protons dissipating at high speed to leave space virtually empty. But Eddington has argued that matter and energy are necessary to extend space, that empty space devoid of geodesics shrinks to a point. So without this delicate balance there would be no cosmos in any intelligible sense.

The Planck time is 10^{-43} sec. Once this amount of time had elapsed after the Big Bang, current theory holds, the enormous amount of available energy produced elementary particles spontaneously, both matter and anti-matter. Among these were so-called X-particles, made up of two u-quarks. X-particles are ephemeral, and each decays into a positron and an anti-d-quark. They are also subject to both strong and weak interactions, so that in the extremely small fraction of a second succeeding the Planck time, they maintained an exact balance between particles and anti-particles. But particles and anti-particles annihilate each other, and unless something had occurred to upset the symmetry between matter and anti-matter, there would have been no matter in the universe, but only the gamma rays in which pair annihilation results. After 10^{-35} sec., however, X-particles and their matching anti-particles decayed into quarks and leptons, and their equivalent anti-quarks and anti-leptons (though not at the same rate), producing a very slight preponderance of matter over anti-matter (one extra quark in every ten thousand million). Subsequently all X-particles disappeared, being normally very short-lived. To this minute imbalance we owe the existence of the material universe, and without it there would have been nothing but undetectable photons (undetectable because there would have been nothing with which they could interact, and so no detectors).

The negative electric charge on the electron exactly equals the positive electric charge on the proton, in spite of their great difference in mass. In every atom there are equal numbers of protons and electrons, so that every atom is electrically neutral. Were this not so, if either the charge on the electron or that on the proton were stronger than the other, however marginally, all atoms would be either positively charged or negatively charged, whichever were the greater. In that case, every atom would repel every other and no macroscopic bodies could ever be formed. All matter would be indefinitely dissipated, and the inner stability of the atom itself would be disturbed. Hence, once more, the possibility of a material universe depends on a precise adjustment, in this instance, of the opposite charges on complementary particles.

Our space-time world has four dimensions, three spacelike and one timelike. There is no apparent inevitability that this should be the case, for mathematicians and physicists conceive of so-called configuration (or phase) spaces of many more dimensions, and are often constrained to do so for the accurate calculation of physical states and quantities. Nevertheless, it has been mathematically demonstrated that the discovered laws of physics are possible *only* in four dimensional space-time.

The existence and nature of all physical phenomena appear to depend upon the propagation of classical and quantum waves, and the properties of wave equations are closely dependent on dimensionality.[2] In less than three spatial dimensions, waves could propagate at arbitrary speeds, whereas in three they are restricted to one only (the velocity of light, c). Consequently, in a lesser number of dimensions signals radiated at different times could arrive at the same point simultaneously, and no sharply defined signals could be guaranteed. Further, it has been shown that reverberation-free impulses cannot be obtained in an even number of spatial dimensions, and that only in three can wave propagation be achieved without distortion. It has also been proved that three dimensions are essential for information-processing of any kind, without which most life processes, and in particular the functioning of the human brain, are not feasible. These facts are not only of the utmost importance for life, which is vitally dependent on high-fidelity wave propagation and the generation and transmission of information, but affect all physical phenomena that involve waves, as all appear to do. A physical universe, properly so-called, could hardly be possible in other than four spatio-temporal dimensions.

The above considerations are further consolidated by the demonstration that the fundamental natural units of the universe, constructed from the primary physical constants (G, h, and c) do not occur except in four dimensions, and upon these virtually all the known laws of physics depend. In any world of other than four dimensions, planetary orbits, whether determined by Newtonian or by Einsteinian laws, would be unstable, making the necessary conditions for the emergence of life impossible. Even more radical, only in four dimensions could the electronic orbits in the atom remain stable, so that matter as we know it can exist.

Not surprisingly, therefore, Ehrenfest (in 1917) argued that all the fundamental laws of physics depend upon the four-dimensionality of space-time, and Hermann Weyl (in 1922) showed that it was indispensable to Maxwell's theory.[3] It should follow that, except in 3 + 1 dimensions, the known laws of physics cannot hold. The superstring theory, indeed, requires ten dimensions, but as six of them are confined to a volume within a radius of the Planck unit of length (10^{-33} cm.), only four are phenomenally discernible. It would seem, then, that only in a space-time manifold of four dimensions is an ordered universe possible.

One might object that a possible world without physical laws could be conceived, and even if it were total chaos, it would be possible notwithstanding. That would, however, strain the meaning of possibility intolerably; for we have already seen that pure chaos is a self-destructive notion, chaos being always parasitic on order of some kind, and order, it seems, is possible only in a world of four dimensions. But, setting that argument aside, even supposing that the fundamental constants were stochastic and statistical in nature, arising only in local or temporary circumstances, Barrow and Tipler point out that only in a relatively cool universe could symmetries be distinguished from chaos, and only at relatively low temperatures could life exist.[4] In other words, the adjectives "orderly" and "chaotic" would be inapplicable to any physical world other than the one we observe, which is the only one compatible with the existence of life and observation.

The constants of nature are dimensionless numbers. Physicists cannot yet deduce them *a priori* from any universal law, although that is one of their theoretical goals. These constants are, however, fundamental to all known physical laws and they have some curious features. There is an unexplained coincidence between certain very large numbers that suggests an underlying, fundamental principle. The ratio of the electric and gravitational forces between a proton and an electron is approximately 10^{40}. The approximate number of nucleons in the universe is 10^{80}. The ratio of the action of the universe to Planck's quantum of action is in the vicinity of 10^{120}. But even more remarkable, the minutest difference in any one of the fundamental constants would alter the whole structure of the physical world so as to eliminate any possibility of the emergence of life. For example, a variation in the strength of the force of gravity by one part in 10^{40} would eliminate from the universe all main-sequence stars, leaving only blue giants (if the constant were larger) or red dwarfs (if it were smaller). As we have seen, only planets revolving round main-sequence stars can support life.[5]

Grand unified and superstring theories have brought physicists to suspect that only one account can be given of the nature of the physical world that provides the necessary symmetries and avoids incoherent anomalies, and the above considerations lend further credence to the view that there can be but one possible physical world, because the minutest divergence from the primary principles and the initial conditions that have determined its overall character would result in self-annihilation. But if there is only one possible physical system, and if it is so constructed as to be delicately adapted to the needs of life, we must presume some essential connection between the generation of living beings and the fundamental nature of the physical world.

INTERDEPENDENCIES

Whether a star belongs to the main sequence depends upon its size, which must fall between quite narrow limits. Stars exceeding a certain threshold of mass become blue giants, while those below a certain limit become red dwarfs. The sizes and masses are determined by several factors all dependent upon the fundamental constants and the precise relations between them. The formation of stars, in the first instance, follows the hierarchical fragmentation of gaseous clouds, and the mass of the fragments depends on the rate of cooling and that of gravitational contraction. The star must not cool too fast to prevent internal radiation from slowing down the gravitational collapse until enough heat can be generated within the cloud to produce equilibrium between the two forces and create a stable star. This delicate balancing act is what ensures that most stars belong to the main sequence. But it depends on precise relations between the gravitational constant, electromagnetic ratios, and thermonuclear forces.

Again, the lifetime of a star is connected with its size and luminosity (the rate at which it burns hydrogen), all of which, as follows from what has just been said, depend upon the fundamental constants of nature. And to them, likewise, the age and size of the universe are related, as well as the time span necessary for the evolution of life forms.

Regardless of other considerations, the sizes of all material objects are determined by the constants of nature. The size as well as the properties of all atomic and molecular systems are regulated by two in particular: the fine structure constant and the ratio of electron to proton masses. In the atom, the centripetal force keeping the electron(s) in orbit round the nucleus is electromagnetic attraction. What defines the characteristic atomic size is the limit of the kinetic energy of the electron at a very small radius. If the atom is small it tends to be stable because the energy is insufficient to generate electron-positron pairs; but in larger atoms there is susceptibility to such pair production and consequent fission. So an upper limit is set to atomic size and weight.[6]

The stability of the atom also depends upon the Planck constant, h, which restricts the electrons to definite quantized orbits, while Pauli's Principle of Exclusion, forbidding two electrons with the same quantum numbers to occupy the same orbit, governs the atomic structure and decides the valency of the atom, determining the chemical properties of the element of which it is an atom. These determinants have far-reaching effects for the synthesis and properties of macromolecules essential if metabolism is to be possible and living processes are to function.

The size of planets depends on equilibrium between gravitation and the electron pressure generated by the effects of the Exclusion Principle, along with electrostatic repulsion. In very large planets the augmentation of gravitational pressure generates heat, which at a certain level becomes sufficient to initiate fusion between hydrogen nuclei with the emission of energy on a scale that would transform the planet into a star. Thus an upper limit to planetary size is established.

The importance for life of the size of the planet it inhabits follows from the fact that only planets above a certain size exercise sufficient gravitational pull to retain an atmosphere containing vital gases, such as oxygen, carbon dioxide, and nitrogen. On the other hand, planets above a certain mass are liable to exert so strong an attraction that they would create tidal effects in living bodies too great for comfort or survival. Further, large planets generate heat (as has been noted) and temperatures exceeding those which life can tolerate.

The viability of organisms also depends on their size, which prescribes, independently of other circumstances, the range and limits of their activities. Three-dimensionality requires that volume and weight increase as the cube, while surface area increases only as the square of linear measurements. This places limits on the functions of the skin of an organism in relation to its bulk.

Barrow and Tipler point out that breakage along a two-dimensional surface is sufficient to fracture the bonds between the molecules of a body.[7] Accordingly, animals should not be so big that acceleration due to gravity is sufficient to cause such fracture should they fall. Considerations like this limit the size of two-legged animals and explain why larger beasts are four-footed (reducing strain on the skeleton and the risk of falling). Such factors as these dictate that elephants have relatively broader feet than smaller animals and stronger and thicker bones. For the same reasons, weight in water being less than in air, aquatic animals can safely become much larger than those living on land.

In the same way, purely physical considerations determine the maximum size and wingspan possible for birds, and such facts as that small mammals lose heat more rapidly than large ones and need more food in proportion to their body size to maintain body temperature, that only very small aquatic creatures can take advantage of ciliatic propulsion, and that unicellular creatures can exist within relatively frail and tenuous membranes.[8]

That biological facts should be traceable back to physical relationships and laws, and ultimately to the constants of nature, should occasion no surprise. Living bodies are material and are bound to obey physical laws. Physico-chemical explanations of biological processes are now common-place and requisite, and (for reasons to which we shall return and have already in part noticed) do not in the least derogate from teleological interpretation. What is important to stress is the necessary interlinkage of the physical with the biological, the unity of the physico-biotic whole, and the continuity of the organizing principle at successive levels of existence.

THE CONGENIAL ENVIRONMENT

Some reasons have already been mentioned why the probability is low that life could exist elsewhere than on earth. Such considerations are strengthened when attention is paid to the extraordinary concurrence of terrestrial circumstances that favour the sustenance of life on this planet.

The planet on which life finds a habitat must, as we have seen, be large enough to retain an atmosphere. It must also have cooled down to a sufficient degree; but that alone will not suffice. It must also be within range of the energy poured out from its parent star, near enough for the supply and warmth to be adequate, but not so close as to make the heat intolerable. Its orbit must be approximately circular rather than elliptical, so that the difference between temperatures at perihelion and aphelion is not too extreme. And to ensure a stable orbit of this shape, the parent star must be sufficiently distant from others to eliminate the danger of collision, or near approach, between stars, which would disturb or destroy the planetary orbits.

Even so, more is required: the planet's rate of revolution upon its axis must be just right to prevent excessive heating in one hemisphere and excessive cooling in the other, as occurs on the moon. Moreover, night and day temperatures must not contrast to excess. Sudden and extreme changes must be moderated so as to prevent excessively violent disturbances in its atmosphere—conditions that can all be fulfilled (as they all are on this earth) only if various concomitants are available.

The atmosphere is needed, not only to provide important gases for the synthesis of organic substances and the sustenance of metabolic processes essential to life, but primarily to mediate and regulate the sun's heat. Ozone in the outer atmosphere acts as a screen against excessive ultraviolet radiation; carbon dioxide vapour creates a greenhouse effect, and water vapour moderates temperatures (because water has a high specific heat) by cooling when they rise too high and warming when they fall too suddenly. Thus the size of the earth ensuring a gravitational pull sufficient to retain an atmosphere, its distance from the sun, its rate of rotation, and the constituent gases of its atmospheric mantle are all nicely adapted to afford the facilities of a living habitat.

In addition to the atmosphere, the hydrosphere is equally indispensable to life. Water is an exceptional substance with exceptional properties, most of which, compared with other substances, are anomalous. It is liquid at the average temperature of the earth's surface, but it freezes, vaporizes, and boils at higher temperatures than other liquids. Its specific heat is higher than most organic compounds; its surface tension is unusually great. It is unique in having a solid phase of lesser density than its liquid phase, so that ice floats, leaving warmer water as a possible refuge for living beings underneath, and providing a buffer against the colder air above. Moreover, if ice sank to the bottom of lakes and seas it could not be melted during the summer and would continually accumulate.

The dielectric constant of water is unusually high, making it an especially good solvent. Its molecules form what is called a solvation cage round the non-polar molecules of dissolved substances, an effect that regulates the shapes of organic molecules like those of enzymes, whose catalytic properties in life processes are due to the precise shapes of their molecules and are specifically diverse in accordance with those shapes.

These exceptional properties of water are due to what is known as the hydrogen bond in the atomic structure of the H_2O molecule. The electronegative atom of oxygen attracts the electron of the hydrogen atom, leaving the positive proton unattended and able to attract the electro-negative atoms in other molecules. This chemical binding force is found in few other compounds and in none that are liquid within the temperature range beneficial to life. The hydrogen bond is what gives water its high melting and boiling points (more energy being necessary to break the molecular bonding), its high specific heat and latent heat, its higher density in the liquid than in the solid form, and its special properties as a solvent.

Water is the dominating constituent of living matter. Its solvent capacity is of inestimable importance for life. It provides the immediate external and internal environment of most living beings. The cooling effect of its evaporation moderates the heat of more extreme climates and enables plants and animals to tolerate higher sun temperatures than they otherwise could. Water vapour in the air buffers the temperature changes between night and day and assists in the greenhouse effect that generally regulates atmospheric temperature. Water is indispensable to life. And the planet earth is unique in the solar system in having a hydrosphere—what may exist elsewhere is still to be discovered.

Water is composed of oxygen and hydrogen, and the hydrogen bond accounts for its special properties. Hydrogen atoms are the lightest and smallest in the universe and the gas is in many ways singular. Because of its lightness it is highly mobile, and it combines readily with many other elements. It is light enough for most of it to escape the earth's gravity, but some of this loss may be compensated by inflowing hydrogen from the sun; and, because of its high chemical activity, it enters into more compounds than any other element. Consequently, in combination with heavier substances it remains in considerable quantities in the terrestrial environment.

The importance of hydrogen for life rests mainly in its presence in hydrocarbons and carbohydrates which are typical components of organic structure. Hydrogen (Barrow and Tipler tell us[9]) forms strong and stable covalent bonds with carbon, which do not easily permit hydrogen bonds with other elements and are the chief source of non-polar side-groups in biological molecules, giving rise to the hydrophilic effect in water.

Life can exist without oxygen, but only in very primitive, anaerobic forms. Higher species cannot do without it. It is the most abundant element not just in living organisms, but also in the earth's outer crust. Of the atmosphere just over one-fifth is oxygen, but this fact has a curious connection with the development of life. At one time, oxygen was scarce in the earth's atmosphere, and its present abundance is the result of photosynthetic activity by plants. Animals burn it up to form carbon dioxide, as do other forms of combustion. There is thus a delicate balance in the atmosphere maintaining oxygen at about 21 percent. If there were more, vegetation would become more easily inflammable and would succumb to fires started by lightning and other causes. If there were less, other inhospitable conditions would arise (for example, an increase in the greenhouse effect, raising atmospheric temperatures to excess). The significance of this equilibrium between oxygen and carbon dioxide is something to which we shall have occasion to return. Meanwhile, it should be noticed that the plentiful supply of oxygen in the earth's crust is another of the special features of the planet.

Oxygen performs another vital function in the form of ozone, which, by collecting in the upper atmosphere, forms a filter that prevents excessive ultraviolet radiation from penetrating to the earth's surface. Ultraviolet radiation destroys carbon-based life, and without the protection of the ozone layer, only creatures living in water, which has a similar screening effect, would be safe.

We speak of life as carbon-based, indicating the paramount importance of another element, carbon. This too is special in many respects. It has a valency of four and can form double bonds with oxygen, giving carbon dioxide (CO_2). As the valency of oxygen is two, the L shell of the oxygen atom is filled up in this combination, so that the carbon dioxide molecule, which has little affinity left for other CO_2 molecules, remains gaseous at normal atmospheric temperatures. In this it is unique among oxides of its congeners in the periodic table. The oxide of silicon, for example, is the exceptionally hard crystal, quartz.

Carbon is slow to react chemically and its compounds are unusually stable. In living organisms, therefore, the organic compounds formed with carbon remain stable until a new combination is required, when it can take place through the catalysis of specific enzymes. Carbon can form over two thousand hydrides, and, because of the strong bonds between its atoms, they can be linked together in long chains or in rings, enabling it to enter into numerous complex compounds, with large numbers of polymers with diverse properties.

Because carbon produces so many free radicals, Barrow and Tipler point out, the probability is increased that, in the primordial "soup," some collection of carbon molecules could spontaneously have happened upon a self-reproductive reaction pathway to initiate the generation of living things. And since carbon forms a greater variety of compounds than any other element except hydrogen, its compounds can store more information than others, and the generation and reproduction of information is the essential characteristic of life.

Living organisms derive their carbon from carbon dioxide, which has properties as special and unique as those of the other vital gases we have considered. It is easily soluble in water and has about the same concentration when so dissolved as it has in air, enabling it to transfer from air to water without difficulty and so to effect easy exchange between organisms and their circumambient element. Being gaseous and soluble in water, it is easily removed from the living body, of whose metabolism it constitutes the final waste product. In combination with hydrogen it forms carbonic acid, so that it plays an important role in maintaining the constant pH level required by that metabolism. Once again, we come upon a chemical element with physical and chemical properties nicely adapted to the needs of life.

The atmospheric balance between oxygen and carbon dioxide, the importance of which was noticed above, is due to photosynthesis, and that again depends solely upon the absorption of sunlight by the chlorophyl molecule. But it can absorb light only of one particular yellow colour. As it happens, the sun's light is just the right colour, which it would not be if its temperature were different. There is thus a precise match between the temperature of the sun, determining the colour of its light, and the ability of chlorophyl to absorb it, without which there would be no photosynthesis, so completely indispensable to the existence of life and so necessary to the atmospheric balance of oxygen and carbon dioxide essential to life's support.

Some of the most important building units in the economy of living organisms are the amino acids that make up the polypeptide chains of proteins and are ingredients of cell walls. Others are the nucleic acid constituents of the chromosomes in the cell nucleus. All of these include nitrogen in their composition. Nitrogen is not very active chemically, but it combines relatively easily with hydrogen to form ammonia, the main source of its availability for the biochemistry of organisms. Although scarce in the hydrosphere and the lithosphere, nitrogen in the atmosphere is plentiful, constituting nearly 80 percent of its content. Once more, the terrestrial environment shows its natural suitability for the synthesis of organic compounds and the maintenance of living organisms, in marked contrast to conditions obtaining on other planets in the solar system, even those comparable to the earth in size and distance from the sun.

There are, besides elements important for the maintenance of life such as sulphur and phosphorous, sources of energy which, by intricate systematic use of enzymic catalysis, can be made to release it in small instalments, so that the supply is accessible without disrupting sensitive bonds. Life also makes use of other heavy elements, which, as noticed earlier, become available only through their dispersal throughout the universe by supernova explosions among main-sequence stars. Here, once more, an extraordinary circumstance links the physical nature of the cosmos with the conditions essential to life.

Supernova explosions result in the formation of red giants, stars whose energy comes from the fusion of helium nuclei into carbon nuclei. This, however, can occur only by way of an intermediate step: the fusion of two helium nuclei to form a very special isotope of beryllium that is highly unstable. The necessary reaction is made possible by a unique two-way resonance between the energies of the helium nucleus, the unstable isotope of beryllium, and the resulting carbon. Without this extraordinary coincidence there would be no red giants, no carbon, and no heavier elements in the universe.

The earth is a planet of the right size, orbiting a star of the right kind, enveloped by an atmosphere with the right composition, and with a hydrosphere unique among the planets. It harbours elements and compounds with extraordinary properties, all propitious and most of them indispensable for the propagation and maintenance of life. None of this would be possible except in a universe as old and extensive as the one we observe, or in any but a space-time of four dimensions, or in any other than a physical world governed by the discovered fundamental constants of nature. These conditions are all interdependent: if space-time had not four dimensions, the constants would not be what they are; if neither of these conditions obtained, there could be no wave propagation and no atoms, whether of hydrogen or of any other kind of matter; and without matter or energy space-time would have no extension. On all this the existence of main-sequence stars and their attendant planets depends. Without hydrogen there would be no main-sequence stars, and without red giants no heavy elements. Without any of these there could be neither atmosphere nor hydrosphere, and without them, no life.

Once again, the extraordinary interconnectedness of things is indelibly underlined, and the oneness of the universe is emphasized afresh. The uniqueness of conditions on this planet and the minute mutual adaptation of circumstances cannot fail to impress. It is not inconceivable that a similar convergence of conditions favourable to life, highly improbable though they are, could occur elsewhere in the vast expanses and among the millions upon millions of galactic systems distributed throughout space. But one may not dismiss as insignificant the precise appropriateness for the emergence and persistence of life on earth of the universal physical laws, and of the qualities and interrelations of substances. The unity of the universe and the exact nature of the organizing principle that governs its order and structure are clearly not indifferent to the emergence and the existence of life and mind. Nevertheless, care must be taken not to misapply, or misinterpret, this mutual relevance of the physical order to the appearance of life upon earth, or its bearing on our intelligent awareness of it.

Of all this, nothing is brought about by our ability to discover it. It is not because we are here that the world comes to be so disposed, but rather the opposite. So-called anthropic explanations are misused if they are understood to suggest such a reversal of causal connections. It is because the world is thus ordered, because the terrestrial environment is so precisely suited to the emergence of life and the development of a biosphere, that human beings have evolved and we are able to investigate the conditions of our own being. Our observation and reflection are not the efficient causes of what they reveal to us, although, as will become apparent later, they may well be its final cause.

CONTINUANCE OF THE SCALE

In the last chapter, we traced the scale of physical wholes from waves and elementary particles to atoms and molecules. We can now go somewhat further. Molecules as they grow in size become more complex, and in chemical combination their atoms enter into intricate structures, determined by the number and degree of saturation of their electron shells. These (as we have said) are regulated by Pauli's Principle of Exclusion and the constants involved in the strong and the electromagnetic forces. The complexity of these structures increases with the formation of macromolecules and polypeptide chains. They also combine into lattices and the leptocosm of crystal forms, all based upon and governed by the same principles.

The growth and replication of crystals is recognized as a transition stage from physical and chemical activity to the self-maintaining and self-reproductive processes of life. Especially do liquid crystals, smectic and nematic, display characteristics similar to living processes; and many substances within the living organism assume such forms. Viruses are now recognized as crystals, and in organisms like foraminifera and sponges there are structures that are crystalline.

Two points emerge: crystals are eminently structural wholes, and they come into being by a recognizable complication, or self-enfoldment, of molecular structure. Schrödinger suggested that the chromosome was what he called an aperiodic crystal, related to the periodic type as the composition of a tapestry is related to the repetitive pattern on a conventional wall-paper.[10] So the postulation of a subsequent self-enfoldment is therefore by no means extravagant, of which the helical configuration of the DNA molecule provides clear indication. Meanwhile, the continuous potency of the ordering principle is ensured by the fact that the fundamental constants governing atomic and molecular structure likewise determine the chemical properties of the polymeric substances of the living cell: the ribosomes, the enzymes, and the DNA of the nucleus.

It is not just one or two notable coincidences disclosed among the scientific facts we have been reviewing that should excite our interest, but the ubiquitous convergence of conditions towards what is beneficent to the propagation and support of life. The unity of the physical world seems, as it were, to focus itself in this convergence, as if it were the implication of its intrinsic order from the very start. The minutest divergence from the initial disposition of forces would have rendered the whole concatenation impossible. And if, as contemporary developments in unified gauge and superstring theories portend, there was originally only one force, from which the known four have "frozen out"—if there is only one primary equation from which all physical forms can be deduced, and only one theory that will eliminate all anomalies and provide all the necessary symmetries—then the delicate equilibria and the precise concurrence of factors that precondition the emergence of life must have been implicit from the beginning.

At one time this striking concurrence of conditions and circumstances auspicious for the production of living beings on earth might have been attributed to the intended purpose of a divine creator. The evidence so far disclosed is such that one may well be inclined to entertain that belief, but it is not yet enough to compel it. It is, however, inept to object that everything hitherto observed might yet have a perfectly natural explanation, without the need (as Laplace averred) to resort to the supernatural. Of course it all has a natural explanation. The very interrelationships of the facts inspected constitute a natural explanation, and nothing so far adduced forces us to resort to outside causes. Physicists are as yet unable to explain the coincidence of large numbers, but Dirac's Large Number Hypothesis foreshadows the possibility of doing so. Nobody has yet offered an explanation of the precise values of the fundamental constants, but there is some reason to hope that the development of a new theory, like supergravity or superstring theory, could provide insight into their apparent interconnectedness. The very meaning of explanation is the placing of a phenomenon in a coherent system of ideas and descriptions that will make its occurrence intelligible. Modern science has come a long way towards doing this for the concurrence and convergence of conditions to which attention has been drawn in these pages, and where it has not yet succeeded, there is no reason to declare that it is bound to fail.

It is not the failure of scientists to explain natural facts that should drive us to a belief in the existence of God. That has other, much more cogent, inducement and support. If any proof of the existence of God is to be found, it must be sought in another fashion than by rummaging about in vain pursuit for evidence among natural causes. We should not emulate the Soviet cosmonauts who reported on their return from space that they had found no evidence of God's existence. The issue is one that cannot be pursued here, but it will be addressed in due course and must await its proper place.

The right philosophical conclusion to draw at this stage is neither that we have evidence of deliberate intention on the part of some conscious agent to foster such conditions, nor that the peculiarities of our own capacities for observation and thinking have fashioned them. It should be that the discovered facts reveal an interdependence among things and processes that forbids attempts to explain matters purely by analysis and reduction to detail (necessary though that may be as one aspect of our methodology). We must look to the whole for an understanding of the parts. We must recognize the overwhelming evidence that physics affords of internal relations between facts and objects, and of their mutual determination. We are constrained in consequence to admit the unity and systematic wholeness of the universe and its subjection to a universal principle of order that regulates all these facts and relations.

It will follow that there is an unbroken continuity between the inorganic and the organic. They contrast as opposites, but they are obviously complementary. The latter simply raises the former to a higher power. The influence of the universal is transmitted uninterrupted, through forms of growing complication and self-enfoldment, along a scale of increasing degrees of adequacy in its exemplification, which guarantees that life is the fruition of what is already potentially present in the physical. Its emergence is simply the continuation of an already-evident tendency to build more integral, more versatile, and more self-maintaining wholes.

Notes

1. Cf. Paul Davies, *God and the New Physics*, p. 179.
2. Cf. Barrow and Tipler, *The Anthropic Cosmological Principle*, Ch. 4, pp. 266ff.
3. Ibid., p. 260f.
4. Ibid., p. 257.
5. Paul Davies, *God and the New Physics*, p. 188. Brandon Carter, "Large Number Coincidences and the Anthropic Principle in Cosmology," *Confrontation of Cosmological Theories with Observation*, ed. M. S. Longair (Reidel, Dordrecht, 1974).
6. Barrow and Tipler, *The Anthropic Cosmological Principle*, pp. 295–297.
7. Ibid., p. 311f.
8. Ibid., pp. 110–118.
9. Ibid., p. 543.
10. Cf. E. Schrödinger, *What Is Life? and Other Scientific Essays* (Doubleday, New York, 1956).

5
Organism

BEYOND PHYSICS

The physical world is in contradiction with itself because, on the one hand, it is spread out in space and time, its contents are mutually external and its events are successive; yet, on the other hand, it is an integral whole, all parts of which are interconnected and interwoven, all elements internally related and interdependent. Its systematic wholeness is governed by a single principle of order, but this manifests itself in the mutual adaptation of disparate elements that register its influence, while they are not apprised of its nature, or otherwise cognizant of each other. Its causation is blind, automatic, in some respects fortuitous, and at times stochastic. Physicists sometimes speak metaphorically of particles "feeling" this or that force, but their expression cannot be taken literally. The unity of the physical universe becomes apparent to us as we study it, but that is the fruit of our observation, our inference and our interpretation; it is not apprehended by the physical reality merely as such. The physical world is a single whole in itself, but it is extended in space and time, its parts and elements mutually external, even though mutually influencing and internally related. As purely physical, its unity is merely implicit, the ordering principle immanent in its elements; it is not explicit for itself and self-reflective. As purely physical, then, its wholeness is incomplete and its unifying principle is inadequately self-manifest.

As a system, however, the universe cannot be only partially whole, for every partial whole presupposes the complete totality; and as the principle of order is immanent in every element and phase of its development, that principle is dynamic and fuels a persistent *nisus* towards more adequate self-manifestation. Consequently, physico-chemical processes tend to become more complex and self-enfolded and to drive towards greater self-determination, bringing the whole into more concentrated centreity and sharper focus within each entity. We have described the physical scale of forms from elementary particles to atoms, molecules, macromolecules, and crystals; the next step is to the living cell, the organism.

HALLMARKS OF ORGANISM

Organism is distinguished from the inorganic primarily by the manner in which the organic being maintains its self-identity. A physical entity is identifiable by the persistence amid changing surroundings of the same particle, or particles in the same relations. In the organism, on the contrary, the material substance which constitutes its body is in constant flux, and its identity depends upon the continuance of its systematic structure and the interrelations of the metabolic processes that both mould and sustain its material form. The principle of organization now manifests itself in a constant dynamic equilibrium among interlaced chemical cycles, automatically adapting and adjusting themselves to changing external conditions.

Some adumbration of this self-maintenance of form within an energetic flux has been suggested even within the purely physical realm. Schrödinger contended that particles were ever repeating *Gestalten* within the wave field,[1] and the superstring theory offers the conception of particles as different vibrations and resonances on a "string" (of what, it leaves to the imagination). But even if we accept these suggestions, we must recognize the organism as the imposition of a new and even more plastic form on a different and yet more complicated substrate process. It is the result of a further and more intricate self-enfoldment of the physico-chemical basis, and, even less than the physical entity, is it a static pattern within this substrate, rather than a continuous movement—a dynamic life-cycle, an open system constantly exchanging matter and energy with its environment.

The processes of life are distinguished from inorganic chemical reactions by a concert of metabolic transformations consititating a system, the effect of which is (within definable limits) to create and sustain an individual structure, spontaneously adjusting itself to external circumstances and adapting its operation amid changing conditions so as to counteract influences that would otherwise bring the sustaining processes to an end. For this capacity of self-maintenance I have in the past employed the term "auturgy."[2] It bestows upon the organism a higher degree of self-reference and individuality, distinguishing it from its inorganic surround. Hans Jonas has described metabolism as "the first form of freedom,"[3] and if one defines freedom correctly as self-determination, this is an apt description. Moreover, the auturgy and response to external changes are the implicit rudiment of that registration of the environing world which foreshadows consciousness. Consequently, with organism the concrete universal has embarked upon a new phase of self-specification, at a higher level of individuation and integrity.

The system of the organism is an open system in dynamic equilibrium with its surround. It is a cybernetic system, which maintains itself in a steady state, or homeostasis, by means of a complex and intricate network of negative feedback, or servo-mechanisms. A negative feedback mechanism is a device which controls the energy supply by diverting a small proportion into a sensor of some kind that feeds information back to the source of the supply, and measures the discrepancy between the present state of the system and a norm somehow established as requisite for the maintenance of the steady state. As the norm is approached or surpassed, the mechanism reduces the supply of energy appropriately, while, if the state of the system declines below the required level, the energy supply is appropriately increased.

No insight into the method of operation of such a cybernetic system can be gained by taking it to pieces and examining the working of its several parts, but only by examining its effect. It can efficiently and appropriately produce that effect only by working as a whole, and its general character can be recognized only by detection of the norm that governs the homeostasis it preserves. Like the organism itself its systematic wholeness, the coherent interaction of its parts, is the sole guarantee of the maintenance of the required steady state, or homeostasis. For this reason, and also because it is governed by a set norm, any cybernetic system must be recognized as teleological.

There are three ways in which the organism maintains itself in the flow of energy and matter upon which it imposes its specific form. These are growth, regeneration of lost or damaged parts and organs, and self-reproduction. Constant renewal of substances essential for metabolism is an obvious necessity for the self-maintenance of an open system in constant commerce with its surround; but a number of limitations restrict mere accumulation (which is, in any case, never characteristic of living organisms). The stability and cohesion of the organism will be threatened by sheer accumulation of substance, so that, in order to survive, it will need either to differentiate into parts with specialized and complementary functions, or to divide itself into daughter organisms, or both. In fact, all organisms adopt both methods, though in different ways, according to the degree of their complexity and development. It is also an obvious necessity for self-maintenance that damaged parts should be repaired and that lost organs should be regenerated.[4]

The definition of life that is being offered here is, then: an open system of chemical processes in dynamic equilibrium capable of maintaining its specific form by spontaneous (auturgic) adaptation to environing conditions.

This way of characterizing life is compatible with the definition accepted by Barrow and Tipler in terms of sufficient conditions: a system containing information preserved by natural selection, which is capable of self-reproduction.[5] Information is equivalent to form. It is technically defined as negative entropy (or order) and corresponds, in general, to structure or organization.

That the organism is a system with a specific form means that it contains information. That it is able, by spontaneous adjustment to surrounding conditions, to maintain that form involves an ability to grow by synthesizing new material, to regenerate and to reproduce. The role of natural selection is conditional upon these prior characteristics, for without specific form there would be nothing to select, without spontaneous adaptation the system would not be alive (an inorganic crystal capable of self-reproduction does not adapt itself spontaneously to unfavourable circumstances), and without both of these there could be no natural selection.

For natural selection to operate, reproduction must already be occurring, and this reproduction may not be simple repetition but must involve some variation of form that will be "selected" only if it gives the organism an advantage for survival over other forms. Moreover, the selection, when it occurs, is not independent (as we shall find) of spontaneous adaptation. There is much more to be said on this matter, which will come in what is to follow.

So far as it is a dynamic equilibrium in a constantly changing medium, the organism is an open system, but so far as it is self-contained and self-complete, it is closed. It is a whole in itself, and the control of its internal processes, the functional regulation of its parts (or organs) by the structure and the organic nature of the whole, is even more apparent than in the case of physical entities like atoms and molecules, the integrity of which is largely dictated from without by general laws and pervasive forces (gravity, electromagnetism, and the like). Within the organism, structure and function are merely two aspects of a single process, constructive of the organ in the course of its functioning. The function, meanwhile, is always and only contributory to the maintenance and integrity of the system. Consequently, it is the system or whole that determines the nature and activity of the parts, and not *vice versa*, and it is this that constitutes the self-determination (or freedom) of the organism and distinguishes it, as a successor and further degree of wholeness to any simply physico-chemical entity, as well as the precursor of consciousness and freely acting personality.

This is not, of course, to say that physical laws cease to apply, or physical forces to operate in the organism. On the contrary, they are essential to its self-maintenance and are used in its auturgy to preserve its dynamic equilibrium and adjust its size to the efficiency of its functioning and its metabolic chemical processes to the exigencies of changing environmental conditions.

THE ORIGIN OF LIFE

How organic wholes of this kind originally arose within an entirely inorganic environment remains an unsolved mystery. No light is shed on it by the hypothesis that spores may have reached the earth from outer space, for the origin of the spores still remains unexplained. Biochemists speculate that cosmic rays or lightning playing upon a primaeval "soup" of carbohydrates and hydrocarbons, aromatic substances, amino acids, and the like, which, it is presumed, could have been synthesized abiogenically, might have effected synthesis of yet more complex macromolecules, similar to those of the proteins and nucleic acids requisite for metabolic processes. A type of protein has been synthesized in the laboratory, but not the sort typical of, or necessary to, living metabolism.

Proteins of the latter kind are, according to present knowledge, only produced within the complex systems of living organisms by stepwise discharges of energy that build them up in successive stages. This method of procedure is necessary because the amount of energy needed to synthesize the entire polymer is so great that, if it were discharged all at once, it would disrupt the process and destroy the substances contributory to its action. The stepwise reaction is effected by the breakdown of high-energy compounds, such as ATP (adenosine triphosphate), and depends on the systematic disposition of specific enzymes acting in a definite order. The synthesis of all these reagents is subject to the same conditions and presupposes the already existing system of the living organism. It seems as if synthesis of the requisite organic compounds cannot be presumed prior to the emergence of the living system itself.

Even if organic compounds, such as those to which reference has been made, could have been synthesized abiogenically and then utilized by primitive organisms (however they might have been produced), the supply would soon have become exhausted unless the primitive organisms were able to synthesize them for themselves. But that involves chemo- or photo-synthesis, both of which are highly complex processes requiring special enzymes and so presupposing the already existing living system. Autocatalysis of any sort is not known to occur independently of the contributory cooperation of specific enzymes reacting in complicated cycles. So whichever way one tries to conceive the generation of metabolism within an environment devoid of life seems to involve some kind of *hysteron proteron*.

The alternative hypothesis, due originally to A. I. Oparin,[6] is that the operation of natural selection upon chain reactions in open systems, occurring at different rates in multiple complex coacervates (colloidal complexes), could eventually result in a self-reproductive system. But this hypothesis fails for a similar reason. Natural selection can operate only upon an already reproducing creature, whose progeny differ from their progenitor by slight variations sufficient to give them an advantage over other competing phenotypes for survival and further reproduction. These prerequisites are not present in a non-auturgic coacervate that does not adapt spontaneously to potentially hostile conditions, and would simply be extinguished if such conditions occurred. Once a self-reproducing auturgic system has come into existence, life is already present, so it cannot be the product of natural selection, to which it is necessarily prior.

Whether, or however, these difficulties may be overcome, what they highlight, once again, is the inherent wholeness of the organism, which cannot be assembled by additive conglomeration of previously independent parts and processes. The whole is *ab initio* prior to the parts, and the nature and interrelation of the parts are explicable only by reference to the systematic structure of the whole.

The problem, however, is not necessarily insoluble, for the primacy of the whole in relation to its elements has already been established at the physical level; and just as the EPR Paradox may be resoluble in terms of non-local faster than light influences, just as the electron in the atom leaps, without apparent transition, from one orbit to another (with the emission or absorption of radiant energy), so possibly some such instantaneous non-local factor may be discovered to account for the sudden appearance of living entities in an otherwise non-living environment. At a critical point in chemical evolution, some threshold in the concrescence of convoluted polymers may have occurred in the primaeval seas, which spontaneously initiated the generation and fecundation of living bacteria. Moreover, the indications of continuity in the degrees of complexity and integrity from purely physico-chemical to organic wholes is impressive, and the presumption remains unshaken that it is unbroken.[7]

WHOLE AND PART

Examples of this dominance of the whole over the part and its direction of sustaining processes are ubiquitous in the biological sphere. Outstanding among them, in embryology, is the regulation and ordering of chemical processes to fit a specific and regularly reiterated design. This is effected in different ways at different levels and in different species. In so-called mosaic eggs the cytoplasm at an early stage is highly fluid; then, by a process defying the law of thermodynamics that prescribes uniform mixture of liquids by diffusion, different substances are segregated into a definite pattern. When the segregation is complete the cytoplasm suddenly becomes viscous, fixing the positions of the segregated substances. Thereafter, segmentation separates in different cells what is needed for the development of different tissues as the anatomy of the adult creature requires. The differentiation of the component substances into the right pattern and the exact timing of the viscosity when that has occurred, neither before it is complete nor after the liquid might have rediffused, is clear evidence of the determination of the detailed processes by the configuration (both spatial and temporal) of the whole. In other forms, this segregation takes place at a later stage of development. At first the cells of the early segmentations are more or less uniform in content and character, but if separated they prove to be equipotential in their capacity each to develop a complete and perfect embryo. In contrast, if left in conjunction, they will multiply and diversify, becoming specialized to produce mutually complementary organs. The same is true of every cell of the cambium (the cylindrical structure running through the stem of flowering plants) in spermatophytes, each of which can generate either root or branch as required.

The regeneration of lost limbs is another case in point, examples of which are plentiful. If the lens of a newt's eye is surgically removed it is regenerated from the rim of the iris, quite differently from its normal embryonic development, which is from the skin opposite to the rudimentary retina. But the final product functions normally, in accordance with the requirements of the organic system.

If flatworms are cut into pieces, each fragment will regenerate new heads, or tails, or both, as required to reconstitute the whole, reorganizing their material and rebuilding the properly proportioned creature. The ascidian *Clavellina*, if mutilated, displays an even more radical self-reconstruction. It degenerates to a uniform mass of cells and then reorganizes itself to reconstruct a new and complete, though smaller, ascidian.

The healing of wounds is effected by the spontaneous division and migration of normally fixed and sedentary cells, which cover the wound with fresh membrane and reconstruct the damaged or missing tissue by processes sensitively adapted to the requirements of changing conditions in the environment, both internal and external. The final result is the restoration of the damaged part in conformity with the needs, the structure, and the functioning of the whole.

As remarked above, living growth is never the mere accumulation of material, but is always nicely regulated to fit the requirements of the total system. Size and proportion are accurately adjusted to ensure efficient functioning of the relevant organs. For instance, in the experimental transplantation of eyes among amphibians, if the eye is given to a host of the same species, but younger in age than the donor, it is found to grow more slowly until the disproportion is rectified between the organ and the rest of the body. A similar adjustment is made in the opposite sense, the organ growing more quickly if the host is older than the donor. It was noted earlier that limbs in larger animals are proportionately thicker and their bones are stronger than in smaller beasts, because weight increases in proportion to volume while the strength of its support varies with the area of the cross-section. The growth rate of the limbs is therefore different in different parts of the body to ensure the correct relationships. The same meticulously adjusted variation occurs in the growth rate of the antlers of deer, which are shed in the winter and regenerate in the spring. As the stag increases in age and weight, the growth rate of the antlers increases each year, but gradually slows down as the weight of the antlers reaches the correct relation to the strength of the animal. Even more spectacular is the growth of the osculum chimney in the young sponge, which is disproportionally large in its earliest stage because the relative paucity of its collar cells prevents it from ejecting water far enough to ensure that it will not suck the same water in again. As the sponge grows, the number of its collar cells increases with its size and the vent pipe becomes inconspicuous in proportion.

Complete physico-chemical descriptions of all, or most, of such processes have been given or suggested. But what most needs explanation is their integration into and regulation by the totality of the system. J. T. Bonner remarks that "there is a general theory to cover every aspect of the control of growth save one, and that is the problem of the *configuration within the whole organism*" (his italics).[8]

Nowadays it is often maintained that the problem has been solved by the discovery and interpretability of the genetic code. But that certainly cannot explain all regulation (in particular, equipotentiality), because the genetic code is reduplicated identically in every cell, so that it cannot account for the cell's ability to develop differently in different situations, sometimes producing a complete organism, sometimes only a specialized part. Again, it has been suggested that regulated development is controlled by something in the organism analogous to a computer programme, but the source of this presumed programme remains totally obscure.

Programming a computer normally presupposes a human agent, who constructs the programme with a definite purpose in mind, but the suggestion that the genetic programme had been planned deliberately by God would hardly be welcome to those who oppose organicism and espouse the more materialistic approach of biochemistry. In principle, it may be said, self-reproducing, self-programming computers are conceivable. But even they would initially require a human (or divine?) inventor and programmer. To contend that such genetic machines could have been evolved by random mutation and natural selection would beg the question because selection can only operate on a self-regulating organism already in existence. Moreover, as I shall argue in the next chapter, it is virtually inconceivable how such a machine as a computer of this degree of complication, let alone any self-regulating programme, could have evolved from unregulated chemical processes through a series of accidental changes, however selected. Finally, nobody has yet identified, or can plausibly identify, anything in the organism that corresponds to a computer programme, short of the human brain itself, and even that analogy is highly suspect and perilously circular.[9]

But even if this problem could be solved in physico-chemical terms (and there is good reason for denying that it could in principle), that would in no way detract from the teleological character of the processes, so long as teleology is understood (in the way prescribed in Chapter 2, above), as it should be, in terms of the domination and determination of the parts by the whole, and not, of course, in terms of any conscious choice on the part of the organism, or deliberate dispensation by some external agency (divine or other), least of all the presumption of the last for the sake of human purposes.

Elaborate feedback mechanisms within the organism maintain homeostases wherever required for self-maintenance and stability, but they are always variable relevant to the demands of environmental conditions.[10] Here again, it is invariably found that what sets the norm for homeostatic regulation and governs the feedback relations is the health of the organism, the maintenance of its integrity and wholeness. To argue that a feedback mechanism is not teleological is perverse, because it requires a setting and a norm serving the needs of the organism (as its name implies), a norm to be met if the metabolic, or physiological, process is not to hamper and disrupt the organic system.

Rupert Sheldrake has listed a number of difficulties facing any attempt to give a purely physico-chemical explanation of morphological phenomena, even with the most optimistic expectations of future discoveries. Description in physico-chemical terms may well be possible, but explanation is another matter. First, he points out, the use of DNA and its propensity to synthesize proteins to explain morphogenetic processes is severely limited by the fact that, in species as different as chimpanzees and humans, the proteins and DNA sequences differ very slightly (less than in different species of mice, or of fruit fly). Secondly, different patterns of development occur within the same organism on the basis of the same DNA; for instance, limbs as different as an arm and a leg, with all their precise arrangements and connections of bones, tendons, and muscles. Thirdly, how are the physical and chemical factors determining differentiation (even if they can be discovered) themselves ordered? Sheldrake cites the example of the slime mould, an aggregation of free-living ameboid cells, which after moving about for a time like a slug grows up into a stalk bearing one or more sporangia. It has been found that the aggregation of the cells is due to the action of adenosine monophosphate (cyclic AMP); but what cannot be determined is whether the distribution of this chemical is the cause or the effect of the pattern of pre-stalk differentiation among the cells. Nor, in any case, could the cyclic AMP account for its own distribution within the slime mould, or for the fact that the structural pattern varies from one species to another. In higher plants, a hormone, auxin, serves some function in the control of vascular differentiation; but then again, the production and distribution of auxin is found itself to depend on the vascular cells, which release the hormone as they mature. Even if these difficulties could be overcome, there is still the problem of regulation to normality of frag-mented parts of a system that has been disrupted or truncated (as in the cases described above). Finally, there are questions of how the polypeptide chains of proteins, once synthesized, fold up into their characteristic three-dimensional shapes, and how they give to the cells their characteristic structures.[11]

These and many other problems stand in the way of physico-chemical explanation of morphogenetic processes. They do not prove that with further research and increasing knowledge more complete chemical des-criptions of the processes could not be given. But they are all of the same kind inasmuch as they arise from the assumption that *explanation* of the whole can be derived from description of the parts; whereas the over-whelming evidence shows that the nature and diachronic development of the parts are determined and directed by the pattern of the whole. This patent fact is what impressed the vitalists who wrote earlier in this century, like Hans Driesch. What he called "entelechy," the determining influence in the development and functioning of the organism, he explicitly identified as its unity. But vitalism failed because it appealed to a mysterious cause which, in fact, had no explanatory power. So organicism followed, intro-ducing (or reintroducing) the concept of fields, developmental, or mor-phogenetic.

The idea of the morphogenetic field has been developed in detail and with considerable sophistication by Rupert Sheldrake.[12] Having persuasively argued the impossibility of giving a satisfactory physico-chemical explana-tion of biological morphogenesis, he proposes the hypothesis of formative causation by morphogenetic fields. This is not energetic (i.e., efficient) causation, but is the non-energetic cause of form (rather than of motion); just as the architect's blueprint is the cause of the form and structure of a building, though not an energetic cause, the energetic causation being provided by the work of the builders and their machines.

Sheldrake applies his hypothesis to physical and chemical structures as well as to biological, maintaining that the difference is only one of degree in the rapidity and complexity of formation. Chemical morphogenesis he describes as aggregative, biological as both aggregative and transformative. The difference is explained in terms of "morphic units"—units (or unities) of form. Chemical and biological systems, he says, are hierarchies of morphic units (e.g., subatomic particles, atoms, molecules, crystals, and so forth), and morphogenesis begins from some already existent centre of organization, called a "morphogenetic germ," around which a higher-level morphic unit comes into being (as, for example, an atom comes into being around an atomic nucleus), either by aggregation of lower-level units (as in physical and chemical structures), or, when the germ is part of a different higher-level morphic unit, transformative, the transformation being induced by the new morphogenetic field. The distinction is somewhat obscure, but need not trouble us unduly at this stage because there are more important and more fruitful aspects of the hypothesis to be explored.

The field is said to be a probability structure, determining the probable disposition of the parts of the developing system. And, as the morphic units are hierarchical in structure, the higher-level morphic unit acts on those of its (lower-level) parts as a higher-level probability structure on lower levels, which are modified in accordance with it. The higher-level morphogenetic field restricts the probabilities of the lower levels. Accordingly, it would seem, not only are the structures hierarchical, but also the formative causation. Thus, in the case of developing structures, such as organisms, the higher-level morphic units (the system in its more complex forms) will exercise a directive influence upon the simpler and more germinal forms. In short, the causation is *final*, the efficient causation being provided, presumably, by physico-chemical activity.

Sheldrake draws attention to the probabilistic character of phase transitions at the physico-chemical level and mentions cases in which "random fluctuations can give rise to spatial patterns."[13] He then remarks that in the living cell far more complex chemical systems are encountered than in the inorganic realm, involving numerous different phase transitions, and he suggests that the probabilities of structure which result could be determined by formative causation—that is, by the morphogenetic field.

What is significant in all this, and particularly relevant to the argument I am developing, is that the appeal to the notion of field, as it has already done in physics, gives priority in explanation to the whole over the part, in opposition to reductionism. Field is primarily a structural concept, a formative whole to which the notion of force, or energy, is subordinated. Similarly, in General Relativity Theory, motion is no longer explained in terms of forces, but rather in terms of geodesics within the metrical field (space-time). Einstein geometrized mechanics, and similarly, in effect, Sheldrake's hypothesis seeks to geometrize morphogenetics.

It may very well be that the appropriate type of geometry is that discovered by the scientists of Chaos, who have examined complex dynamic systems, finding them to be ordered by "strange attractors" and to exemplify fractal curves. They have discovered how turbulence generates stable shapes, and how phase transitions may be represented as the boundaries between fractal basins.[14] Sheldrake himself mentions in passing the work of R. Thom, who represents the final form of a morphogenetic process as an "attractor" within a morphogenetic field.[15] In short, one might suspect that the strange attractor, with the forms it generates (of which the Mandelbrot set is the compendium), is the same thing as the morphogenetic field. Biological forms were, in fact, just what these scientists succeeded in generating in their computer graphics.[16]

What has to be admitted, surely, is that the organizing principle of the whole is not a physical or chemical force but a different kind of effect altogether. That is why Pierre Teilhard de Chardin, in his remarkable book, *The Phenomenon of Man*, called the integrating power that produces new and more complex systems "radial energy" and distinguished it sharply from physical (or what he termed "tangential") energy. So Sheldrake distinguishes formative from energetic causation. It is significant that what are known in physics as saturation forces, arising from the operation of Pauli's Principle of Exclusion, are not strictly dynamic forces at all, but that the principle functions as an organizing effect structuring the orbits of electrons in the atom and determining, in consequence, the valencies and chemical properties of elements and the leptocosm of crystals.[17] Is not this just formative causality at work?—the regulation of formative processes by the principle of organization governing the whole.

DIALECTICAL RELATIONSHIPS

Although the organism is in persistent and intimate commerce with its physico-chemical surroundings and is wholly dependent on them for its supply of energy and matter, it is also in continual conflict with them, because it is an open system in dynamic and delicate equilibrium that is easily upset by external changes and is constantly under threat of destruction from influences discordant with its internal organization. Consequently, the organism wages a perpetual struggle with its inorganic environment, by repeated alterations and adjustments of its internal processes, in order to maintain its complex homeostases and its organized integrity.

Accordingly, organism and environment are at one and the same time mutually continuous and mutually complementary, while also in opposition and conflict. Further, the living whole represents a higher degree of complexity and integration than the inorganic and simply physico-chemical substances of which it is ultimately made up. Its metabolic processes are a further self-enfoldment of the inorganic chemistry that is nevertheless indispensable to and constitutive of them, and its structure is a multiple complexification of molecular and crystalline forms. It is thus a continuation of the dialectical scale already traced in the sequence of physical and chemical structures. Life relates to the non-living as opposite, and equally as complement, and also as a new and higher degree of specification of the universal organizing principle of reality, now manifesting its unity in its own active and auturgic self-maintenance at a higher level of self-sufficiency and individual self-determination. It is the next stage, after the physical and the chemical, in the dialectical scale of forms, into which the universal whole differentiates itself in its persistent drive towards coherent self-completion.

Notes

1. Cf. E. Schrödinger, *What Is Life? and Other Scientific Essays*, p. 177.
2. Cf. Errol E. Harris, *The Foundations of Metaphysics in Science*, pp. 180 et seq.
3. Hans Jonas, *The Phenomenon of Life: Toward a Philosophical Biology* (Harper and Row, New York, 1966), p. 3.
4. Cf. Errol E. Harris, *The Foundations of Metaphysics in Science*, Ch. X.
5. Cf. Barrow and Tipler, *Anthropic Cosmological Principle*, p. 515.
6. Cf. A. I. Oparin, *The Origin of Life* (Oliver and Boyd, London, 1957).
7. Many of the points made above have been discussed in more detail in *The Foundations of Metaphysics in Science*, Ch. IX.
8. J. T. Bonner, *Morphogenesis* (Princeton University Press, Princeton, NJ, 1952; Atheneum Paperbacks, 1963), p. 147.
9. Cf. Rupert Sheldrake, *A New Science of Life: The Hypothesis of Formative Causation* (Collins, Paladin, London, 1987), p. 26, and Roger Penrose, *The Emperor's New Mind*, Chs. 9 and 10.
10. Cf. Errol E. Harris, *The Foundations of Metaphysics in Science*, Ch. XI.
11. Cf. Rupert Sheldrake, *A New Science of Life*.
12. Ibid., Ch. 4 and passim.
13. Ibid., p. 88.
14. Cf. Ch. 3, pp. 33–36 above, and James Gleick, *Chaos: Making a New Science*, pp. 235–236.
15. Cf. Rupert Sheldrake, *A New Science of Life*, p. 60.
16. Cf. James Gleick, *Chaos*, pp. 108ff and 238–239.
17. Cf. Henry Margenau, *The Nature of Physical Reality* (McGraw Hill, New York, 1950), Ch. XX, and "The Exclusion Principle and Its Philosophical Importance," *Philosophy of Science* 11 (1944).

6
Evolution

THE *SINE QUA NON* OF NATURAL SELECTION

The unit of life is the cell, but it is in itself a highly complex structure, differentiated into proteins and nucleic acids, cytoplasm and nucleus, and containing specialized organelles, Golgi bodies, ribosomes, and mitochondria. More than that, it is an unceasing activity, a constant movement of concerted and concatenated cycles of chemical analysis and synthesis. It is thus no mere unit but rather a microcosm of life—a conception that we shall find later to be more apt than that of a simple building block. Sir Charles Sherrington, in his Gifford Lectures, has described it as follows:[1]

> The cell is a unit-life . . . The cell is not a polyphasic chemico-physical system merely. Many a mere drop of complex jelly could be that. The cell is a polyphasic chemical system which is integratively organized. Hence there comes about that it can answer to what is described as 'life' . . . It is dynamic. It is energy-cycles, suites of oxidation and reduction, concatenation of ferment-actions . . . We seem to watch battalions of specific catalysts, like Maxwell's 'demons', lined up, each waiting, stop-watch in hand, for its moment to play the part assigned to it. Yet each step is understandable chemistry.

Although each step is understandable chemistry, the integrative organization is not chemistry at all, but must be explained, if it ever can be, in terms of some holistic principle. It might be referred to the genetic code in the DNA of the nucleus, but whence this genetic code originates, or how, nobody yet knows. It is itself an organized system the source and principle of which has still to be discovered.

Moreover, to be alive the cell must auturgically modify its internal activity and external behaviour to adapt to the local conditions in which it finds itself. The protozoan *Paramoecium*, if the water in which it lives is deprived of oxygen, moves away and adheres to a bubble where the oxygen supply is richer. Similarly, the rhizopod *Arcella*, in like circumstances, secretes a bubble that floats it up nearer the surface where there is more oxygen. *Stentor*, a vase-shaped unicellular animal, fixes itself by its apex to a solid surface and feeds through the orifice at its wider open end, which is supplied with cilia to waft in edible material. If bombarded with inedible particles, *Stentor* first sways about in evident endeavour to avoid the bombardment, then reverses the wafting action of its oral cilia to drive the particles away, and if that fails, shrinks into a blob, detaches itself from the surface to which it had been anchored and swims away. Another protozoan, *Stylonychia*, if starved shrinks in size and progressively rearranges its substance to maintain its internal organization (which is exceptionally complex for protozoa), maintaining the appropriate relations and correct proportions between the parts.

Behaviour of this kind characterizes the auturgic self-maintenance of living things; and that, as we have seen, involves growth, regeneration, and reproduction. On nothing less than unicellular creatures, like those above described, can natural selection operate. Only auturgically adaptive, self-maintaining, self-reproductive metabolic systems rank as living, and nothing less can be "selected" for survival in competition with other such living reproductive beings. Any non-auturgic, non-adaptive chemical system (for instance, Oparin's complex coacervates), if it encountered conditions inimical to its structure, would simply disintegrate. Only if it could adapt to changing conditions and reproduce itself might its progeny, when it differed from its parentage in ways that favoured its adaptive capacity, acquire selective advantage over competitors for survival and then be "selected" from among competing forms. Adaptation is inherent in the very nature of life and cannot originally have depended on natural selection, because, without it, natural selection could never have begun to operate. This point has to be made before any sensible discussion of evolution can begin, for if it is overlooked, serious misconceptions are liable to arise.

The form in which life originated may have been unicellular, or protistans may have separated out from colonial masses of cells which first acquired living characteristics. In either case, something must at some stage have favoured the survival of colonial aggregations. But none of these are mere collections of protistans in symbiosis. At some stage in their life cycles, the cells of such colonies differentiate, as when the slime mould *Dictyostelium* develops from a slug-like mass of ameboid cells into a delicately stalked sporangium, some cells being specialized to form the stem and others to become spores. *Chondromyces crocatus* goes through an even more elaborate process of differentiation in its fruiting stage. More complex colonial forms, like *Obelia*, reproduce both by budding and by the production, through separate blastostyles, of new free-swimming medusae. Presumably, the mutual support of the colonial cells must give such forms "survival value," and their capacity to differentiate, however acquired, must provide some added advantage. Eventually the differentiation results in the separation of somatic from generative cells, and sexual reproduction appears, giving ampler scope for natural selection, so that evolution can proceed apace.

Progression from relative uniformity to differentiation and specialization of organs is an advance in the degree of complexity and organization, marking another step forward in the scale of forms. Metazoa represent a fuller expression of the organizing principle than do protozoa (with corresponding relations between unicellular and multicellular plants); and the effect of further evolution is to extend the scale still further. But before we can consider the significance of evolution in this regard we must examine its nature, its course, and its so-called mechanism somewhat more closely.

CHANCE AND DESIGN

The accumulated evidence is so massive that living species have evolved from progenitors of different genotypes, and are still evolving, that the theory is now firmly established and its truth is automatically assumed by respectable biologists. No other theory of the origin of species (e.g., creationism) has scientific support. Likewise, the evidence is ubiquitous and copious that evolution is promoted by chance mutation and natural selection. But the current neo-Darwinian assumption that this is a sufficient, or even a necessary, condition of its occurrence may be seriously questioned.

Variations occur from a variety of causes; some are heritable and others not, and those that are heritable result from alteration of the genetic material in the gamete. Mutations in a gene will produce heritable changes in the phenotype, which, if they are not immediately lethal, may give the mutant an advantage for survival over other forms, either in the prevailing environmental conditions, or in changing circumstances less favourable to the former varieties. In such cases, the mutation is said to have "survival value." If it is disadvantageous, the mutant will eventually be eliminated from the population in competition with the unchanged members of its own species, and the transmuted gene will be bred out. These general facts are established beyond reasonable doubt. But there are weighty reasons for believing that they are not the whole story and that there are other important factors contributing to the evolutionary results.

First, the phrase "natural selection" is not always used with due care. That a variant with even a slight advantage for survival under prevailing conditions should thrive and proliferate is natural enough, but that in no way implies a selecting agency. Domestic breeders actively and deliberately select the parents of the varieties they desire to breed, and it was this process that gave Darwin the idea that environmental pressures might have a similar effect in nature. But, in the latter case, no deliberate choice is involved. The new variant is not literally selected, it simply survives, and less suitable forms are naturally eliminated. It is very important to remember that the *only* effect of "natural selection" is elimination of the unfit. It makes no positive contribution to the fitness of the mutant, accidental or other, nor does it set any norm to which the mutating variety must conform in order to survive. The defect of some recent attempts to show by computer modelling how pure chance variation could produce specific types is the surreptitious (and no doubt unconscious) introduction by the experimenter of some such norm.

The negative effect of natural selection was emphasized by T. H. Morgan and L. T. Hogben some fifty years ago,[2] when they remarked that if (*per impossibile*) there had never been any selection, all the known forms of life would still have appeared, as well, no doubt, as innumerable others which would have escaped extinction. Accordingly, the present results of evolution owe nothing to natural selection (with some slight qualification presently to be noticed), which has only had the effect of eliminating the less viable forms that would otherwise have persisted.

If the current neo-Darwinism is espoused, we must assume that all the minute, delicate, and immensely complicated servo-mechanisms, and the cunningly devised life-styles adapting modern organisms to the complex conditions of their habitats, are the product solely of chance mutation and random shuffling of genes (due to the crossing-over of chromosomes and similar movements in the genetic material). The probability that this could have occurred within the time-span of terrestrial evolution would undoubtedly be so small as to make the hypothesis scarcely credible.

Julian Huxley, himself a professed neo-Darwinian, once wrote: "To produce such adapted types by chance recombination in the absence of selection would require a total assemblage of organisms that would more than fill the universe and overrun astronomical time."[3] He also calculated (after Muller) that the probability of a successive accumulation of chance favourable mutations sufficient to produce an animal such as the horse would be of the order of one in a thousand to the millionth power.[4] That, he admits, is equivalent to impossibility.

Barrow and Tipler give the odds against the spontaneous assemblage (by which they presumably mean the chance concatenation) of the human genome as falling between $10^{-12,000,000}$ and $10^{-24,000,000}$. *Homo sapiens*, they point out, requires 110,000 genes to code the necessary enzymes and structural proteins, the odds against the assemblage of any one of which are unimaginably vast, between 4.3×10^{-109} and 1.8×10^{-217}.[5]

The negative effect of natural selection, however, applies only to the occurrence of disadvantageous variations. It has a positive effect in making successful varieties predominant in the population, so that fresh mutations will occur, for the most part, in forms already best adapted to the environment. If such mutations then bring further advantages, the resulting phenotypes are likely to be more complex and more subtly adjusted to changing conditions. This effect tends towards orthogonal progression, which is not, however, universal in the course of evolution but is restricted to separate branches. Nevertheless, natural selection has no influence on the actual occurrence of mutations, and none whatsoever on the likelihood of their being beneficial or harmful to the organism.

The great majority of chance mutations are unfavourable, and many are lethal at the outset. Some that would be favourable if they occurred in the right sequence would be harmful if they occurred singly in the wrong order. They would then be bred out, and once that had happened they could scarcely ever recur. It follows that the improbability of producing the extraordinarily complex and intricately adjusted organisms that exist today solely through the accretion of accumulating chance variations is astronomically great. So great, in fact, that the currently assumed age of the earth would be woefully insufficient to allow for its accomplishment. Small wonder, then, that C. H. Waddington should have declaimed: "To suppose that the evolution of the wonderfully adapted biological mechanisms has depended on selection of a haphazard set of variations, each produced by blind chance, is like suggesting that if we went on throwing bricks together into heaps we should eventually be able to choose ourselves the most desirable house."[6]

In *The Foundations of Metaphysics in Science*, I marshalled what I believed (and still do) to be the most cogent and wide-ranging evidence available against the doctrine that evolution comes about solely through random mutations and natural selection, and on that evidence I based the strongest arguments I could muster.[7] The evidence has never been questioned, nor have the arguments been refuted, so far as I am aware; so I can only assume that they still stand. I know no way of improving upon them now, and can do no better than refer the reader once again to my earlier writing.

There, in the first place, I tried to rebut the response of the neo-Darwinians to what Bergson called "the argument from improbability," by showing that, although complex organs, like the human or the cephalopod eye, could be shown to have developed in stages, from a pigment spot to a light-concentrating, image-forming photoreceptor, this could not have been simply by successive additions of fortuitous and random variations, because such variations would be favourable, even when of the right sort, only if they occurred in the right order, and only if modifications in the structure and function of other organs, required to ensure that the developing eye worked more efficiently, were made concomitantly. It is not just that the eye is a highly complex organ, but that its effective use is not possible without the coordinated functioning of associated muscles, glands, neural engrams and behavioral dispositions, involving numerous other organs and parts of the body (for instance, the reflex turning of the head to bring and to keep moving objects in focus). If all these factors were to be supplied piecemeal, by chance mutations, they must occur in the proper sequence and mutual association, which is not only stupendously improbable, but, if the mutations occurred in the wrong order, they would be disadvantageous and selection would eliminate them.

I drew attention to the fantastic complexity of the human brain, with its 10^{10} nerve cells and their bewilderingly intricate interconnected networks of dendrites and axons, and I asked how we could envisage the assemblage of such an organ simply by random and dissociated modifications, all of which (it must be borne in mind) have to be directly related to the integration and control of the activity of other parts of the body.

There are three types of interrelationships to account for, to do which the neo-Darwinian theory must be hard put. The first is the complex and precise interplay of servo-mechanisms in the organism. Consider the supply of insulin: it must be in the right quantities and at the appropriate rate to maintain the correct level of blood sugar. Any mutation affecting one of these factors would be detrimental unless it also modified the other in a complementary way. It would then, if not alone sufficient to kill the individual, be selected against and be bred out of the species. This is one of the simpler examples. The operation of the sympathetic nervous system is far more complex, involving perceptual and judgemental dispositions as well as glandular and neural functioning in numerous and intimately interrelated convolutions. No such closely interwoven and intricately organized system could be the result of merely additive, haphazard, accumulation of elements. The assumption that it could be overlooks the necessity, and evident presence in its operation, of a dynamic organizing principle.

The second correlation that has to be explained, and that must sorely tax the ingenuity of the neo-Darwinist, is the development of behavioural dispositions to fit in with protective colouration. The latter is easily attributed to chance variation and natural selection; but how then is one to account for the accompanying and reinforcing behaviour that makes the protective colouring effective? For example, the longitudinal markings on the wings of the moth *Venusia veniculata* match the appearance of the veins of the leaves on which it settles at night. But if the markings are wrongly oriented, the camouflage is destroyed, as it would also be if the moth extended its antennae. The insect, however, carefully orients its body so that perpendicular markings on its wings match those of the leaf, and tucks away its antennae under its wings. If disturbed, it readjusts its position with obviously directive movements. Other similar examples are plentiful, and none of them can easily be disposed of in terms of random mutation. Once the beneficial modification has occurred, natural selection will ensure its persistence and predominance among the population. But unless it occurs at the right time and concomitantly with the supporting modifications of the system, the effect of natural selection will be precisely the opposite. In that case, the potentially beneficial mutation will have been bred out and lost, probably irretrievably.

The use and adaptation of organs for toilet (the rough tongue of the cat, the teeth of the lemur, and the like) seem to have no obvious relation with their natural functions. The lemur uses its teeth to comb out its fur, and the base of its tongue is abnormally enlarged so as to be more effective for cleaning the teeth. If such concordant developments are to be explained as the result of chance variations, these must be coordinated so that habits of behaviour are suited to the formation of the organs used, and *vice versa*. How long must evolution wait for this to occur through random muta-tions? And if it did, how accurate and precise would be the adjustment of structure to function? More important, how great is the chance that mean-while the mutation would not be bred out?

Thirdly, there are the cases that involve cooperative behaviour between several different individuals, as in the disguise of the nest of the Ceylon black-backed shrike, where not only are the colour and type of the materials for nest-building selectively chosen to match the background and location of the nest, but also the behaviour of the chicks (whose natural colour is similarly protective) is instinctively adapted to make the nest look like a broken branch. Here instinct, colour, and social behaviour are all involved, and if they did not converge to the same end would severally be detrimen-tal, and would be eliminated by natural selection.[8]

The force of the argument I am pressing is not that chance variation and natural selection play no part in evolution. Far from it. That they occur ubiquitously is copiously supported by the evidence. It is simply that they are not sufficient to account for the results, which are not additive accumu-lations of characteristics but intimately and integrally organized systems of structure and function. The moral is that a further, more organismic, account of evolution is required, and that nothing less will do justice to the observed facts.

ORGANISMIC GENETICS

Evidence is available in plenty for such an account of genetics, much of it provided by the very scientists who profess to be neo-Darwinians.

In the first place, as we have insisted, natural selection cannot even begin to operate unless a dynamic, living, auturgically self-maintaining system is already at work. Adaptation and adjustment to environmental conditions is inherent in the very nature of life. To be selected, a system must be better adapted for survival than others with which it competes for the available energy and sustenance. On the one hand, the simpler and less ordered the system, the more probable will be its physico-chemical occurrence in nature; on the other, the living system, by increasing the complexity of its metabolism and its structure, increases its capacity to adapt to prevailing conditions. By so doing it increases its chances of survival and will therefore naturally be selected. Thus, what is selected is the more improbable, the system that has the spontaneous capacity to decrease entropy and increase information. As this goes counter to thermodynamic trends, it requires for its explanation some *nisus* towards greater order and complexity, some organismic principle.

Geneticists have established that single genes do not control or determine single characteristics, but that the chromosome functions as a whole, as does, in fact, the entire genome. There can thus be no question of a simple summation of initially disconnected characters, each displaying selective advantage. Genes are pleiotropic, having multiple effects, and characters usually involve not one but several genes. Moreover, there is mutual influence among the genes, so that their position in the chromosome affects what they produce. Consequently, this will be modified by translocation and the crossing over of chromosomes. Accordingly, mutant genes are "buffered" by their internal environment, by the effects upon them of other genes and their position in the chromosome. Huxley admits that gene combinations adjust the mutant to "the needs of the organism,"[9] implying that there is a *nisus* towards self-maintenance and holism characteristic of all life. In short, mutants, so far as is feasible, are *integrated* into the system not merely collated haphazardly, and then, if beneficial, they are preserved by natural selection. There is obviously an endogenous *nisus* to the maintenance and integration of the organic system, and what is selected is the more efficiently self-maintaining and self-determining whole.

It is now apparent that "survival value" is equivalent to more efficient self-maintenance and more completely self-determining wholeness. Any variation in the hereditary mechanism, however produced, whether by chance mutation or other causes (and chance mutation has not, by any means, been established as the sole cause of hereditary change), that enhances the organism's self-sufficiency and auturgic capacity to adapt to prevailing conditions, will have survival value and so will naturally be selected. What evolves is always the organic system and nothing less; and what evolution produces is increased self-determining adaptation, increased capacity for relevant variation and selective reaction to circumstances, in short, increased versatility and freedom.

A whole with these characteristics is a more adequate manifestation of the self-specifying universal expressing itself in the organism as well as in the cosmos, than is any inorganic purely physical or chemical whole. In biotic forms an advance has been made beyond the merely physico-chemical without abandoning physics and chemistry or ceasing to depend upon its laws. The advance consists in the emergence of new and more self-sufficient wholes, reproducing themselves in ways which result in increasing integrity, complexity, and versatility in adapting and fitting into the environment (a capacity that will presently be seen in a new light and with fresh significance). Organic systems of this kind more fully reflect the nature of the principle of organization immanent in life and in the universe as such, and approach more nearly its free self-determination.

THE DEPLOYMENT OF EVOLUTIVE LIFE

It is a fairly safe assumption that life began with the single cell, whether it occurred in isolation in the primaeval ocean or (as Teilhard de Chardin conjectured) appeared suddenly and spontaneously at one unique instant, in pervasive profusion, as the cohesion of polymeric substances in the ferments of the primaeval seas reached a crucial threshold.[10] In either case, Oparin's speculation of multiple complex coacervates becoming self-reproductive and auturgic is not improbable.[11] At all events, single-celled life was doubtless the original form.

As the simplest and most natural form of reproduction is cell division, the primordial living creatures proliferated by segmentation; and it is highly probable that at some stage daughter cells remained in contact. Thus colonial types (even if not original) would eventually arise. The next stage would be progressive specialization of function among the cells and their mutual adjustment, so as to produce the same effects that the metabolism of the single cell accomplishes, but with greater precision and efficiency. Thus metazoa must gradually have evolved. The specialization and separation of somatic and genetic cells, then, marks a critical advance, introducing sexual reproduction and giving scope for greater variation on which natural selection can work with more spectacular results, moving evolution (so to speak) into higher gear.

But even before this stage had been reached, life had expanded, evolving and ramifying like a growing tree, the trunk dividing and subsequently redividing to form three great branches, first distinguishing plant from animal and then, among the latter, vertebrate from invertebrate. Along each line the forms, in classes, orders, genera, and species, increase in complexity and elaboration, by continually ascending steps, from fungi to phanerogams, from sponges to lepidoptera, from chordates to mammals. The plants are fundamental and indispensable to all the rest because they alone are capable of photosynthesis, on which all life originally depends, and which generates the oxygen that makes the rest animate (gives them breath), so that all animal life is parasitic in some way on the plant kingdom.

Each branch has specialized in a particular manner, and this specialization limits its evolution within bounds that exclude other significant and complementary lines of development. Plants, owing to their immobility and relative fixity, could not develop the same degree of versatility of reaction to environmental conditions as has been possible for animals; nor, in consequence of their fixed location, have they needed to acquire organs of sensory reception to apprise them of objects at a distance. Accordingly, they are strictly limited in the extent to which they can advance towards conscious apprehension of the milieu to which they belong. That they are not altogether incapable of sensibility is evident from the sensitivity to touch of some plants and the ability of others to trap insects. But it is in animal rather than in plant evolution that we must look for the stages through which life has progressed towards bringing to consciousness the whole in which it inheres—the cosmos of which it is itself a differentiating phase.

The protistan is a whole in itself, yet is incessantly dependent upon and in commerce with its surrounding medium. Organism and environment are opposed yet complementary, while the active initiative resides in the organism. To fuel its auturgy more efficiently it needs to reach out into its surroundings to grasp the nutriments it requires and to expel the waste products of its metabolism that, by their accretion, hamper its functioning. To do all this it has to augment its capacities by reduplicating itself and projecting the different functions of its auturgic activity into associated cells, which thereby become specialized. So it evolves into a metazoan. Protozoa and metazoa are again opposites, yet complementaries, the single cell providing the unity and the multiple organism the diversity of the organic whole. Cilia now become nematocysts (lashing, penetrating, thread-like appendages, stimulated to action tactually by contact with other bodies), then tentacles, to capture prey, or they combine into proto-fins or legs for locomotion. So the organism widens its reach and extends its grasp of its surroundings, and the process is set in motion of inwardizing the circumambient world through sensibility and reaction. It continues with increasing versatility and range as special organs develop for proximal and distal sensation.

The evolutionary drive then branches out into opposite and interdependent lines: plant and animal, invertebrate and vertebrate, aquatic and terrestrial, cold- and warm-blooded, instinctive and intelligent. In no case is the prior form totally abolished, but, while it is transcended and transformed, it is raised to a higher, more labile and more comprehensive power.

Nor is any of these contrasting types independent of the others: as will presently become apparent, they require each other in mutual symbiosis.

Thus, the biosphere differentiates itself, through the process of evolution, into a scale of forms that are mutually opposed, mutually complementary species, genera, orders, and classes (or what Teilhard calls verticils): distinct examples of the universal organic system, in differing and progressively intensifying degrees of integrity and complex unity.

ORTHOGENESIS AND PROGRESS

In stating my case in this way, I may be accused of having begged the questions whether evolution is progressive at all and whether it is orthogenetic, both of which have been hotly disputed. That evolution has not proceeded in a straight line is obvious from the evidence that divergent branches separated off at an early stage, and did so again at repeated intervals, as well as from the patent fact that diverse developed forms have survived. The reputed "higher" species have not had to await the extinction of all the "lower" types, nor have they all evolved in strict sequence one from another. But unless there had been some orthogenesis along specific lines, the entire theory of evolution would be otiose, and we should have to revert to some discredited version of creationism that postulates the original fabrication of fixed species. There is general agreement among biologists that evolution has not proceeded in a straight line, but there is equally widespread agreement that different species have descended from common ancestors, and that orthogenesis must have occurred in ramifying directions.

But is evolution "progressive"? Some have argued that it is not, but is simply the constant change of living forms subject to natural selection under environmental pressures.

It has been alleged that mosquitoes have been more successful in adapting themselves to wide differences of climate than have humans. Such comparisons, however, are unworthy of serious attention. No mosquito is capable of making them. Peter Medawar, in his review of Teilhard de Chardin's book, ridiculed the idea of progress, especially in what Teilhard called cerebralization.[12] But the late Professor Medawar's own achievements stand in telling evidence against his strictures. No creature with less "braininess" (to use his own sarcasm) could have achieved what he did. The fossil record and much other evidence gives testimony that species have appeared in sequence, and show progressively increasing organization of the nervous system, with ganglion complexifications growing in size and concentrating forward in the head. This development is evident in both the invertebrate and the vertebrate lines. It is accompanied by steadily increasing capacities for variable and appropriately directed behaviour (generally describable as rising degrees of intelligence), which culminates in all cases with social orders of specific types. There is no other plausible way of tracing or describing the emergence of intelligent life.

In the invertebrate line, in accordance with the law that specialization restricts possibilities, the development culminates in the impressive organizations of social behaviour among ants and bees; but their specialization confines them to an instinctive rigidity and regimentation of behaviour typical of species that have reached a cul-de-sac in their development. Among vertebrates, enough plasticity and variability have been retained for this limiting rigidity to have been avoided, in particular among mammals and especially among primates. Here the evolution of forefeet into prehensile hands and the adaptation of the hind limbs to bipedal gait relieved the jaws of the function of gripping and carrying and eliminated the need for massive jawbones and jaw muscles. So the cranium was left unencumbered and could become enlarged to accommodate a brain growing in size and organization to regulate and direct new capacities. The erect stance of the animal likewise freed the hands for grasping and manipulation and brought the eyes nearer together in the head for forward vision, concentrating the gaze and making binocular vision possible, with increase of acuity and depth perception. All this greatly enhanced opportunities for sensory-motor learning and facilitated the development of intelligence.

Thus the gamut of living species constitutes a dialectical scale of forms that progressively express with increasing adequacy the principle of order and unity immanent, not only in the organism, but also in the environment—in the biosphere as a whole. This, we shall shortly find, is itself a further explication of the physico-chemical totality, reflecting it by stages in more self-determined and self-sustaining wholes. The scale proceeds towards more intelligent and self-conscious forms by way of the development of yet another aspect of its evolutionary advance, namely, behaviour.

BEHAVIOUR

The self-maintenance of the organism is rooted in metabolism; and while this always remains the case, metabolism develops (as the anatomy of the creature becomes more complex and specialized) into physiological activity. Further, both metabolism and physiological process are continuous with behaviour, directed to the ingestion of food and to reproduction, as is apparent from protozoa onwards. As living forms develop further, added to these objectives more and more is behaviour directed to the protection and care of the young. Reproduction is at first mere partition, then the production and virtual abandonment of eggs to hatch in the watery environment into a free-swimming larval stage, which develops independently of parental attention. Later the parent (male or female) watches over and cares for the eggs (as the male stickleback protects and wafts oxygen-laden water over the nest which he has built, into which he had enticed the female and stimulated her to lay eggs he had then fertilized).

Up to amphibians little more than this is ever undertaken, but as terrestrial species develop further, at the level of birds, parents devote their entire activity to food-gathering, mating, and caring for the offspring. Mammals are the only living things (except for some few species of fish) that retain the embryo in the mother's body and continue after the birth of the young to care for them until they are mature enough to fend for themselves. At the higher levels the final stage includes some type of education and guidance.

As this development approaches its later stages, the behaviour tends to become more and more gregarious (with minor exceptions), as is to be expected when it centres on the family group. As the herd, or flock, grows in size, a degree of order frequently appears within it, and it may be described as social. With *homo sapiens* all this blossoms out into vastly more complex and significant behaviour, and only at the human level is social conduct organized in that distinctive fashion we have come to recognize as political or civilized.

Since the early decades of this century, psychologists have advocated and fostered the belief that the only scientific way in which mental activity can be described is in terms of behaviour. The habit has therefore been encouraged of ignoring the part played by consciousness in animal and human activity. Those who have espoused this belief, one would imagine, should not relish our referring to them as thinkers, although we may suspect that that is how they value themselves. The habit is fortunately declining among contemporary psychologists, whose researches take a more cognitive direction. And indeed, although some defence and excuse may be made for behaviourism, for certain purposes, the project of describing behaviour while ignoring (or even denying) its cognitive aspect is surely perverse, if only because what marks living behaviour as distinct from merely physico-chemical reaction is, at the very least, sensibility, and at higher levels (e.g., that of cephalopods or of fish) perception—the response not merely to a "stimulus" but to a situation that is apprehended as a whole.

Behaviour is the extension, through evolution, of metabolism and physiological process. As the metabolism of *Arcella*, when the oxygen is scarce, leads to the secretion of a bubble, so that of *Paramoecium* prompts it to swim to more aerated water. In homoiothermic animals, homeostatic demands lead to physiological changes to regulate temperature (reducing it by sweating or dilation of capillaries, or raising it by shivering) and then, further, to the animal's movement into the shade or into the sun (as circumstances require). The waving of cilia develops into the lashing of nematocysts, the active grasping of tentacles, and the skilful darting and seizing of prey (as practised by cuttlefish). Thus behaviour is the developed continuation of the auturgic activity of the organism in the varied quest of self-maintenance—the active preservation of dynamic equilibrium and wholeness. As such, it is characterized by relevant variation, and it blossoms in the higher species, into sensory-motor, perceptual, and intelligent learning.

Properly speaking, behaviour is informed activity. It is informed in two senses of that word, structurally organized and perceptually enlightened. Behaviour proper is the response not so much to a stimulus as to a total situation, which must be grasped as a whole if the behaviour is to be appropriate. It is thus the conative counterpart and the outer, bodily aspect of cognitive awareness. To the mental aspect further attention will be given anon under the heading of perception. Something may be added here about the structural form.

Basically, all behaviour is instinctive, and the most recent studies have shown that instinct has a characteristic structure, initiated by an appetitive phase, which may be relatively simple or highly complicated and variable, and which is always purposive in that it pursues a definite goal characteristic of the particular instinct (eating, mating, migrating, nesting, and the like). The aim of appetitive behaviour is to bring the animal into the presence of certain highly specific "sign stimuli," which trigger the release of a more or less automatic (partly reflex) "consummatory act," usually fairly stereotyped and relatively unvarying.

Within this general pattern, the appetitive activity is hierarchically subdivided into ancillary, subordinate structures of the same general form. As described by Tinbergen,[13] the reproductive instinct of the stickleback includes, in succession, fighting behaviour, nest-building behaviour, mating, and care of offspring, each further subdivided and displaying the same characteristic combination of appetitive and consummatory activity. The appetitive behavior characteristic of each phase is variable in response, not just to single sensory stimuli, but relevantly to the situation confronting the animal as it demands reaction appropriate to the aim of the instinct. The stickleback, for instance, having chosen its nesting site (a choice adapted to the nature of the terrain), builds a nest and defends it against other approaching males, by fighting behaviour varied to suit the circumstances and the perceived menace of the intrusion. On the appearance of the female, behaviour is again appropriately varied; and the laying of the eggs prompts new and further varied activity. The instinct is a structured pattern of activity, regulated and directed to a specific aim, that contributes to the welfare of the individual and its offspring and the maintenance of the species as an organized totality.[14]

In the course of evolution, with the development of nervous systems and increased brain capacity, the variability of appetitive behaviour increases and becomes more versatile and adaptable, subject to learning in the course of experience. It has been well established that such learning (despite efforts of the early behaviourists to demonstrate the contrary) involves insight, that is, the apprehension of relations, the appreciation of situational exigencies, and genuine efforts to overcome obstacles in the pursuit of anticipated goals.[15] Such learning clearly involves memory, imagination, and the interpretation of presented objects in the light of past experience. As evolution proceeds, it is the progressive emergence of intelligence.

The inner, mental aspect of instinctive behaviour and its intelligent outcome will be treated below. It belongs to a further phase of the self-differentiation of the universal whole; one that renders it aware of itself and its own relational structures. In speaking of it here I have unavoidably been anticipating—unavoidably because the successive phases are mutually implicated. Behaviour (with increasing degrees of intelligence) is fore-shadowed below the mental level in the living processes of metabolism and physiology which, as they evolve, fold back upon themselves to produce new wholes and more developed forms. My point here is that three aspects of evolutionary progress go together and are inseparably connected: increase in brain capacity, elaboration of instinctive behaviour, and improvement in cognitive discrimination and definition—what we may describe as the degree of comprehension of relational configurations. When the human level is reached, this cognitive aspect attains to the pitch of explicit thought, the principle and agency of organization. The process of evolution thus transpires as one of progressively bringing to explicit self-consciousness the principle of organization inherent in the organism from the start, that which galvanizes its auturgy and is itself the immanent principle ordering the cosmos as a whole. How this is accomplished at the human level will become more apparent as we proceed.

CONCLUSION

To sum up, we may say that the course of evolution has unfolded a complex branching series of forms, consisting of wholes within wholes, systems within systems, organisms within organisms. They are all auturgically self-maintaining and, as the scales proceed, they increase in versatility and ability to appropriate the environment (as well as to adapt to it). At every level, and at each stage, what presents itself is an organic whole: the cell, the protozoan, the colony, the metazoan, each continuous with the next and successively incorporating while transforming that which precedes. Thus stages and species overlap. They relate one to another as opposites, yet they are mutually continuous and interdependent, so they are mutually complementary. Each embodies and exemplifies, in its specific degree, the same principle of organization, while, as the scale proceeds, the form in which that principle specifies itself is a more adequate expression of its character, progressing from metabolic self-adaptation, through increasing degrees of physiological and then behavioural efficiency, to conscious (perceptive) appreciation of the presented situation. The entire scale, moreover, reveals itself as a harmony as well as a diatonic sequence, as a single all-inclusive whole, the biosphere, the next subject for investigation.

Notes

1. Sir Charles Sherrington, *Man on His Nature* (Cambridge University Press, Cambridge, 1951), pp. 66 and 70.
2. Cf. T. H. Morgan, *The Scientific Basis of Evolution* (Faber and Faber, London, 1932), p. 130, and L. T. Hogben, *The Nature of Living Matter* (Routledge and Kegan Paul, London, 1931), p. 181.
3. Julian Huxley, *Evolution, the Modern Synthesis* (Harper, London and New York, 1942), p. 475.
4. Julian Huxley, *Evolution in Action* (Harper, New York, 1953), pp. 41–42.
5. Barrow and Tipler, *The Anthropic Cosmological Principle*, p. 565.
6. C. H. Waddington, *The Strategy of the Genes* (G. Allen and Unwin, London, 1957), pp. 6–7.
7. Errol E. Harris, *The Foundations of Metaphysics in Science*, Ch. XII, Sec. 3.
8. Further examples, some more telling and compulsive, are given in *The Foundations of Metaphysics in Science*, Ch. XII, Sec. 3.
9. Cf. J. Huxley, *Evolution, the Modern Synthesis*, p. 48.
10. Cf. Teilhard de Chardin, *The Phenomenon of Man*, Book II, Ch. 1. The same idea is put forward in Hegel's *Naturphilosophie* (see above, p. 44 and p. 45, note 14).
11. Cf. A. I. Oparin, *The Origin of Life*.
12. Peter Medawar, "Critical Notice on *The Phenomenon of Man* by Pierre Teilhard de Chardin," *Mind* LXX (Jan. 1961).
13. Cf. N. Tinbergen, *A Study of Instinct* (Oxford University Press, Oxford, 1952).
14. Sheldrake perspicaciously includes behaviour as subject to regulation by the morphogenetic field. See *A New Science of Life*, Ch. 11.
15. Cf. W. H. Thorpe, *Learning and Instinct in Animals* (Methuen, London, 1956, 1966), Chs. V and VI.

7
Biosphere

ORGANISM AND ENVIRONMENT

In Chapter 4 the close attunement of the terrestrial environment to the needs of life was examined in some detail. From what was there disclosed we may safely conclude that it would be a grave error to imagine evolution to concern only the organism and its changing forms. Life, we have repeatedly affirmed, is a dynamic equilibrium, auturgically maintained between organism and environment, so that there is continual intercourse between the two, and the distinction is merely relative. They form one organic whole and cannot be strictly separated. Evolution is a process involving both together. Plants affect the soil and the atmosphere, as the atmosphere and the soil affect them. Animals do likewise, and each modifies the other, both in nature and in structure. For example, the cropping of verdure by herbivores causes some varieties of plants to flower near their roots, while other varieties of the same species flower normally higher up the stem. Likewise, the teeth of herbivores are adapted to the type of plants on which they graze. The evidence is widespread.

It has been suggested that life originated not in one point but, like the crystallization of a solid in a supersaturated solution, in a single flash, all over the primaeval oceans, wherever the concentration of macromolecular protein-like substances had reached a critical threshold. In this way the hydrosphere would have been laced with a network of interdependent living organisms that would form the matrix of a biosphere. This is pure speculation and is dubious, because there are good reasons for believing that life could have begun only in shallow bays, estuaries, or lagoons. But however that origination did in fact take place, there is no doubt that the individual auturgy of living organisms is only the inner aspect of their commerce and interdependence with their environment, and that this is as much the living environment as the non-living.

All living forms are mutually dependent. The more complex and developed almost universally require bacterial activity for the fermentation and digestion of nourishment, whether at the interface between the root hairs of plants and the soil or in the internal organs of animals. All depend for sustenance on the unbroken continuity of a food chain stretching from plankton to roast beef. Plants provide animals with food and with oxygen. Animals fertilize the soil in which plants grow with phosphates and nitrates while they breathe out carbon dioxide which the plants use in photosynthesis. Carnivores prey upon herbivores, and if they did not herbivores would multiply excessively and exhaust the available pastures. This is only one example of the complex and pervasive interaction by which nature maintains a balance between the species.

LIVING COMMUNITIES

Organic wholeness is not confined to living units. A drop of water can contain a miniature ecosystem, as does every natural pond (uncontaminated by human industry) and every artificial aquarium constructed by human ingenuity. Niches in a forest tree collect moisture, humus, and debris, providing rooting places for epiphytes. The roots of these epiphytic plants, in symbiotic association with fungi, form a network that makes suitable nesting sites for ants. The ants accumulate more food material for the epiphytes and protect themselves (and incidentally the tree and the epiphytes) by stinging larger animals that approach their nests. Holes in the tree's bark collect water where insects can breed whose larvae feed on leaves and other plant material, while the nymphs pollinate the flowers of the tree and of the epiphytes. Birds and small animals also find a habitat in and around the tree, feeding on the insects and fertilizing the surrounding earth. The whole community is a single complex of interdependent life forms in organic solidarity.

But no such ecosystem is altogether self-contained. The tree is linked to its surroundings in the same sort of symbiosis as is involved in its own community, and the entire forest constitutes a similar biocoenosis. This again joins with the adjacent streams and mountain ranges to determine the climate of the area, so that the whole region forms a single system. Nor does the community stop there. What has been said of the forest applies equally to the rest of the continental terrain, and the same biotic network prevails throughout the seas.

Similar spectacular examples of biocoenosis are provided in estuaries where plankton, marine life, riparian flora, and aquatic bird-life are all mutually dependent. Atolls in the Pacific constitute similar communities, as do coral reefs. Marston Bates, in his book *The Forest and the Sea*, has given vivid and fascinating accounts of such ecosystems, illustrating the interdependence of human and animal species on land with vegetable species and marine life.[1] Turtles come on shore to lay their eggs, while spending the rest of their lives in the sea; sea-birds mate and nest on land but forage at sea to find their food; land-crabs depend on both terrestrial and marine conditions, including other species, for their habitat and survival; and countless other interlacing relationships sustain a complex system of symbioses, to which spatial limits cannot be rigidly assigned. The ocean as a whole harbours a single complex biocoenosis that is nevertheless organically related to its surrounding and surrounded land masses, themselves domicile to like communities, and all intimately intertwined with and conditioned by (as well as conditioning) atmospheric circulation and content, tidal and convective movements of the waters, and climatic conditions. In the last resort, the planet as a whole is one ecological totality, changes everywhere affecting conditions everywhere else.

THE ORGANIC EARTH

We said earlier that the living cell was better conceived as a microcosm than as a unit of life. Lewis Thomas, indeed, in his delightful collection of essays, *The Lives of a Cell*, treats the earth as a macrocosm of the cell.[2] He says he can think of no better analogy for the organic integrity of the planet than that of a cell, including within its cytoplasm minor quasi-organisms, organelles, mitochondria, chloroplasts, ribosomes, plastids, and the like. As the cell is enclosed in its protective membrane, which serves as a filter through which simultaneously energy and nutrient liquids are admitted, waste products expelled, and their flow regulated, so the earth is enshrouded in its atmosphere, performing similar functions, regulating temperature, circulating sustaining gases, receiving and dissipating toxic substances. As chemical signals instigate special processes in the cell, so do thermal, chemical, and vibratory signals in the lithosphere, hydrosphere, and atmosphere alert species of living creatures to conditions appropriate for migrating, mating, or spawning. The earth as a whole presents the characteristics of a living being.

Thomas draws these parallels by way of analogy, but J. E. Lovelock proposes, as a scientific proposition, what he calls the "Gaia hypothesis," that the biosphere actually is a living whole cybernetically controlling its earthly environment to maintain the conditions most favourable to its own preservation.[3] The idea, he acknowledges, is as ancient as the Greeks. Although he fails to mention later examples, it was revived and advocated by F. W. J. Schelling in *Die Weltseele* and by G. W. F. Hegel in his *Philosophie der Natur*. Bergson, too, believed that evolving life comprised all reality, the physical world being (at most) its discarded detritus and (at best) the frozen image, or the abstract concept, of an artificially spatialized section through the perpetual flux of *durée*, an abstraction that the human mind constructed as an aid to action which has given it a selective advantage for survival.[4]

Lovelock observes that the fossil record affords clear evidence that since life first appeared on earth, some three and a half thousand million years ago, the climate has changed very little, in spite of considerable variations in the heat of the sun, the surface properties of the earth, and the composition of the atmosphere. This betokens a steady uniformity in atmospheric conditions for which pure chemistry at the inorganic level could not account. To maintain an equable temperature, the greenhouse effect must have been sustained, and this would require replenishment of ammonia, carbon dioxide, and methane as these gases were used up by the growing biosphere. It would also involve some sort of compensatory activity by the biosphere itself, because all explanations involving merely abiological chemistry break down in the face of the needs for survival of early life. Moreover, oxygen and methane react under the influence of sunlight to produce carbon dioxide and water at a rate that would require the replenishment of a thousand million tons of methane and double that amount of oxygen each year to keep their proportion in the atmosphere steady. No known chemical process could do this with any acceptable degree of probability.

The submission that Lovelock makes is that the generation of methane by anaerobic microbes in the mud of estuaries, ponds, and lakes is regulated so as to control the amount of oxygen in the air, preventing it from becoming so great as to threaten general conflagration of forests and grasslands, or so small as to deprive living beings of their energy sources. Contributory to this control of oxygen, Lovelock surmises, is the production of nitrous oxide by living organisms. This gas conveys oxygen to the atmosphere from the soil and the sea bed at twice the rate it is consumed in the oxidation of exposed reducing material, and so may serve to counteract the effect of methane.

> The chemical composition of the atmosphere [Lovelock writes] bears no relation to the expectation of steady-state chemical equilibrium. The presence of methane, nitrous oxide, and even nitrogen in our present oxidizing atmosphere represents violation of the rules of chemistry to be measured in tens of orders of magnitude. Disequilibria on this scale suggest that the atmosphere is not merely a biological product, but more probably a biological construction: not living but like a cat's fur, a bird's feathers, or the paper of a wasp's nest, an extension of the living system designed to maintain a chosen environment.[5]

By like reasoning, he infers that the salinity of the sea must have been regulated throughout the aeons during which life has evolved. If salinity exceeds 6 percent, living cells break down under osmotic pressure. Simple calculation shows that the quantity of salts washed down into the seas from the land masses in each eighty-million-year cycle is equal to the total salt content of the ocean at the present time. For life to have survived, salts must somehow have been leached from the waters, but no known process of ordinary chemistry could have had this effect.

Lovelock suggests that minute shells and skeletons of diatoms and coccoliths, constantly raining down from the surface waters to the depths of the ocean, carry away excess saline material and bury it in the sediments of the sea floor. Increase in salt content of the water would provide more silica and calcium carbonate for the construction of protozoan shells and skeletons and so might promote their proliferation, thus creating a feed-back mechanism to control the overall proportion. Further, the accumulation of chlorates and sulphates from evaporites is greatest in shallow bays and landlocked lagoons, where evaporation is higher and the inflow from the sea mainly one way. So the fact that such lagoons are often created by the activity of living organisms (such as corals) may again mean that life takes part in regulating the saline content of the ocean.

In this theory, the suggestion is also made that the plate tectonics of the earth's crust may also be driven by biological influences. The accumulated weight of biological detritus on the sea floor creates inequalities of pressure that could deform the thin softer rocks at the bottom of the ocean. The blanket of silica would then act as an insulator concentrating the heat conducted from the earth's interior to soften the rock even more and eventually to cause it to melt and give way to internal pressures that explode in volcanic action.

It is not made altogether clear whether we are to conceive all this as accidental concomitance of inorganic chemistry and biotic activity bringing biologically beneficial results, or the product of some inherent pattern of activity in Gaia, the living whole. It could hardly be the consequence of natural selection, for Gaia is not in competition with other planet-sized organisms for scarce nutriments, and is depicted only as manipulating her own inorganic environment. Moreover, she does not produce offspring among whom selection could be made. Accident and selective pressures should therefore be ruled out. For the most part, Lovelock speaks as if the terrestrial organism was just the biosphere, operating upon an atmosphere, a hydrosphere, and a lithosphere as surrounding media. He reminds us that the term "biosphere" was first introduced by the Russian scientist Vladimir Vernadsky, and it was freely used by Teilhard de Chardin, but only to refer to the living envelope of the planet, analogous to its gaseous and liquid envelopes. The evidence offered, however, can leave one in no doubt that biosphere, hydrosphere, atmosphere, and lithosphere are all in intimate organic relationship and interchange, and that they constitute a single organic whole. If the hypotheses of Lovelock and of Sheldrake are taken together, very promising and intriguing consequences emerge.

GAIAN MORPHOGENESIS AND EVOLUTION

Postulating the existence of a morphogenetic field, Sheldrake has raised the question of its origin. He speculates that it might evolve through what he calls morphic resonance, by which an already formative cause persists and could combine with others, so that past structures are reassembled in subsequent morphogenesis. As the resonance is said to be in no way energetic, just how it operates remains somewhat obscure. But, if we espouse the Gaia hypothesis along with that of the morphogenetic field, we shall have to contemplate a field covering the entire planet and directing all the fantastically complex interrelated levels and phases of morphogenesis, with their cybernetically controlled homeostases. It would make the whole earth one organism with an eminent degree of autonomy and self-determination—a freely acting individual.

A field is a configuration of lines of force or causal influence, which may well be a hierarchical structure. My earlier discussion of Sheldrake's theory suggested that formative causation must be so; it operates from the top down, rather than from the bottom up. In effect, it is final, not merely efficient, causation. In that case, a planetary field would be the source of all subordinate fields, and the question of origin would be pushed back to become one of cosmological evolution.

The earth cannot be treated in isolation from the solar system, if only because its own motion and physical state depend almost totally on the sun's gravity and energy outflow. Nor can the solar system be separated from the galaxy, or the galaxy from the rest of the universe. If, as the physicists assure us, the universe is one system and its fundamental laws and forces can be traced back to a single principle, immanent at its origins in the Big Bang, then we must presume that there is ultimately only one universal morphic field.

It would be a universal *morphē*, pervasive over and regulating all subsidiary fields. It would be somewhat akin to Plato's Idea of the Good,[6] the idea of all ideas, shaping all others and determining their interrelations—that which makes all things intelligible, the creating and sustaining principle of all things, the goal of all development or purpose. It will later become apparent that it is also (like Plato's Idea) the source of all truth and the goal of all knowledge. Where Plato's doctrine may fall short (if indeed it does, for pregnant passages in the *Sophist* should not be forgotten[7]) is the impression it leaves that the Idea of the Good is a static, abstract principle, "laid up in heaven," rather than a dynamic, self-realizing *nisus*. We need to move in the direction taken by Aristotle, for whom the form of forms is an *energeia*, an eternal activity, in man "an activity of the soul" and in itself the active reason, *noesis noeseos*.[8]

Sheldrake mentions both Plato's and Aristotle's theories as possible metaphysical options, though he seems somewhat disinclined to adopt either, leaving such ultimate questions entirely open. But if contemporary physical theory and its implications are taken seriously, surely the choice will be rather restricted and very strongly indicated.

The Platonic-Aristotelian notion of a universal form of forms concurs harmoniously with the account I have given of the essential nature of the whole, as an organized system of systems structured in accordance with a universal principle of order, that specifies itself in and as a scale of overlapping, contrasting, but complementary forms, each of which in succession is a provisional whole, manifesting in progressive degrees of adequacy the universal principle of organization. The living cell, the metazoan organism, and the branches and genera of living species range themselves in order displaying these dialectical relationships: invertebrate and vertebrate, aquatic, amphibian, aerial and terrestrial species. And the biosphere relates to the physical world in the same way.

The inorganic and the organic are opposite and complementary, but the biosphere is a self-generating, self-regulating system, expressing the autonomy of the whole more characteristically than the purely physical. Nevertheless, each stage presupposes, depends upon, and is constituted as, a self-enfoldment of the prior stage: the atom, of elementary particles; the molecule, of atoms; the crystal, of molecules, and the polypeptide chain, of radicals; and so on up to the organism and the biosphere *in toto*. Throughout, the universal principle of organization prevails, Pauli's Principle giving the first indication of its effect, and each successive stage revealing more fully its intrinsic nature. As the scale advances, auturgy becomes instinct, and instinct, through learning, becomes intelligence, activity progressively coming to be more freely self-determining and self-aware.

The homeostatic self-regulation of the organism requires a degree of sensitivity to the ambient impinging influences from the outer world, as well as some form of registration of its own inner temper. So it reflects its surroundings and its own relation to them and becomes a focus of the structure of the macrocosm itself. Just as metabolism has been described as the first beginning of freedom, so cybernetic self-regulation may be regarded as the initiation of awareness, both of self and of other.

It is hard to credit the capacity even of the lowliest living forms to adapt and regulate their activity to changing conditions, as they do, without some rudiment of sensibility. In the protistan and lower living forms this registration of inner and outer conditions is (presumably) still obscure and inexplicit; but as organisms become more complex and more highly developed, specialized sense organs develop, more efficient and more delicately attuned to incoming vibrations and contiguous chemistry. So sensation is refined to perception, and the reflection of the surrounding world becomes more articulate.

If now we include in our purview the biosphere as a whole, the nature and course of evolution takes on a new aspect. It can no longer be regarded as just a matter of chance variation and natural selection. Gaia is not engaged in a struggle for survival against competing members of her own or any other species. Evolution now assumes the aspect rather of a process of maturation, in which the development in symbiotic organisms of sense organs and of perception marks, as it were, the way in which Gaia gradually, and by dialectical stages, brings herself to consciousness of her setting within the world and of her own integral unity. More than this, because of the continuity of the process from physico-chemical to biotic, Gaia's bringing of herself to consciousness in the mentality of her member organisms is at the same time the coming to consciousness of the entire cosmos, of which Gaia herself is a specific phase.

This again conforms to the account given above of the nature of the whole, which must in principle and in fact be complete, and which cannot be totally fulfilled unless it is fully and explicitly *for itself* (self-aware). Our next task, then, must be to follow the development of this self-awareness and examine its implications, not only for ourselves as a race, but for the universe as a whole.

The unitary whole that the physicists have discovered the universe to be is now revealed as not simply physical but also alive. The universal organizing principle has specified itself in an extended series of subsidiary and provisional wholes, from elementary particles, atoms, and molecules to viruses and bacteria, to sentient and conscious organisms, each in its own degree expressing the implicit order and exemplifying the totality of which it is a dialectical moment, so that the succession of phases constitutes a graded scale of overlapping, mutually implicating, and interrelated forms issuing as intelligent minds. The major divisions of the series themselves constitute an analogous scale, physical, chemical, and organic, successively continuous, each incorporating its predecessor and each, in its continuous outgrowth from its forebears, dominated by the ordering principle, first manifested in the physical world, then in the biosphere, and subsequently in the noösphere—the realm of self-reflective intelligence, where it becomes aware of itself in explicit conscious knowledge. To a consideration of this latest phase we now must turn.

Notes

1. M. Bates, *The Forest and the Sea* (Random House, New York, 1960; Vintage Press, 1965).
2. Lewis Thomas, *The Lives of a Cell* (Viking Press, New York, 1974; Penguin Books, Harmondsworth, 1987).
3. J. E. Lovelock, *Gaia: A New Look at Life on Earth* (Oxford University Press, Oxford, 1979, 1987). Gaia is the ancient Greek name for the goddess Earth.
4. Cf. Henri Bergson, *l'Evolution Creatrice* (P. Alcan, Paris, 1907; A. Skira, Geneva, 1945), and *Matière et Memoire* (P. Alcan, Paris, 1896; A. Skira, Geneva, 1946).
5. Lovelock, *Gaia*, p. 10.
6. Cf. Plato, *Republic*, VI, 504e–509b.
7. Cf. Plato, *Sophist*, 245d and 247e–249d.
8. Cf. Aristotle, *Metaphysics*, XII, 1074b35, *Nicomachean Ethics*, I and X, *De Anima*, III.

8
Mentality

DEFICIENCY OF LIFE

Life is a prodigously complex system of metabolic processes deployed in space and time, intrinsically physico-chemical yet more automatic than the merely physico-chemical. This autonomy exhibits more adequately the principle of order immanent in the phases of the universal, but deficiently because it is still incognizant of the immanent cybernetics that work in it automatically but not self-consciously. Its wholeness is still implicit, and its centreity, while recognizable by us as observers is not for itself a subject. As yet the unity of the whole is not transparently focused on itself but is diffused in its mutually teleological material functionings. Its parts and organic processes adapt to one another, and the organism as a whole adapts to its environment auturgically, but not yet intentionally. To achieve the capacity for more self-determined action, a higher degree of internality and centralization of external differences is necessary; a further phase transition is required at a new threshold.

As Schelling pointed out in his *Ideas for a Philosophy of Nature*,[1] an organism, as organized being, involves a *concept* because it is a whole constituted by parts that are mutually adapted and are equally adjusted to the overall structure of the whole, each both end and means in relation to the rest and to the whole. Hence, in order to exist at all, such organism *must* be organized, and it can arise only out of what is already organized being. The very functioning of the parts and processes of the organism presupposes (as we have already found) the prior existence of an organized whole. Hence, even the material existence of organism involves a concept (that is, a principle of order and relationship). But a concept implies the existence and activity of a cognizing mind,[2] while the material existence and operation of the organic being is in space and time, dependent on physical laws and external causes that are antithetical to the purely ideal. This contradiction can be resolved only if, on the one hand, the concept immanent in the material system *qua* organized is somehow objectified or actualized in its practical functioning, and, on the other hand, the organizing principle in the organic system is brought to self-consciousness.

103

In yet another way mere organism, as such, is self-contradictory. Although the ambient elements are its indispensable complement and nourishment, to maintain its own organic integrity the organism has to struggle incessantly against environmental pressures. Likewise, it competes unremittingly with members of its own and of other species for limited food supplies and available habitat, yet it is in inseparable symbiotic community with all neighbouring life forms. Consequently, its organic life is in persistent contradiction with itself, a contradiction that the organism is under constant strain to resolve. This it strives to do through an inherent urge to appropriate and internalize its other, to unite and integrate the outer world into its own self. It does so, in the first instance, by absorption and ingestion of material substances required for the maintenance of its bodily integrity, but, in the second place, and in the quest for this sustenance, through sentience, the feeling and registration of external influences and conditions, that marks the next phase transition in the self-differentiation of the universal principle of order, the next step in the scale of forms—what constitutes the matrix of the succeeding phase, mentality.

THE FORM OF THE BODY

The organism is a highly complex system of chemical processes, or metabolism, which combine, break down, and re-combine diverse and exceedingly complex compounds so as to maintain the system auturgically. This metabolism develops, as the processes (uniting function with structure) congeal into bodily organs, to become an intricately organized system of physiological activities that serve and, in their cooperative integration, constitute the health of the individual. Each function is instrumental as a means to the maintenance of the integrity of the system, and the system as a whole is what maintains the aptitude and efficiency of each and all of its organs. The organs are mutually ends and means, as are the different functions and the total system—the traditional identifying character of organism as such. At a critical pitch of intensity, the ordered integration of physiological functioning, as Susanne Langer has said, is felt.[3]

Throughout the course of our argument we have traced wholes in hierarchical progression, and each succeeding whole has brought with it a new form (in the scale of forms) carrying a supervening quality not displayed at previous levels. Thus the atom has a new and more complex structure than any of its constituent subatomic particles, which, to constitute it, do not simply collect into a heap but are systematically organized. In this form, the atom evinces properties and propensities that its forerunners (and now its constituents) in isolation did not possess. The new propensities are peculiar to the new structure or form. It can now unite with other atoms

to form a molecule, which, as a further complexification, is capable of
bonding with other molecules, both in crystalline lattices and in new
compounds with fresh qualities and properties, dependent on the structure
of the molecules involved.

When we reach the stage of macromolecules, their versatility is greatly
enhanced and they enter into the metabolism of organisms in numerous,
intricately interwoven processes as enzymes, vitamins, hormones, and
countless other vitally functioning substances. In every case, the new
emergent properties depend upon the form and pattern of arrangement of
the constituent atoms. The complex whole that appears at this level displays
the emergent quality and the new capacities of life, impossible at any of the
prior stages. Life is the *form* assumed by the integrated metabolic processes.
Now, when these develop and combine as physiological processes, integrated by vascular and neural functioning at a new threshold of intensity, a
further form emerges—namely, sentience.

While form and structure are generally synonymous, cognizance must be
taken of the differing levels in the scale to which the terms apply. The word
"structure" usually has spatial reference and a geometrical connotation, but
in the physical and biological forms listed above, mere spatial pattern is
transcended. Atoms and molecules are energy systems, and it is the form of
the energetic complex that displays the peculiar properties. Life-forms are
intricate cycles of chemical activity, and the capacities of the living organism are the exertion of powers, generated by the systematic integration of
its interlaced chemical sequences. The proposition now being advanced is
that this integration of physiological processes at a high degree of complexity and intensity assumes a new form, the experience of feeling.

This idea, somewhat surprisingly, has aroused perplexity and mystifica-
tion in the minds of some critics.[4] It is easy to recognize the emergence of
new forms at lower levels. Three lines drawn at random on paper have an
arbitrary and indefinite structure, but arranged so as to intersect in three
points they form a triangle, which has a distinct form not possessed by any
of them individually nor all of them taken together in disarray. The form is
clearly recognizable although it is not an additional constituent of the
figure—it is the figure itself, its shape—and it possesses new properties
absent from its separate components.

A collection of atoms milling about in confusion constitute a gas; but
atoms combined systematically as a molecule and cohering in a lattice
determined by the molecular form become a crystalline solid. Once again,
the new character is easily detected, yet it is not a separate, or separable,
component of the complex, while the constituent atoms may be the very
same ones that formed part of the original gaseous diffusion.

In precisely similar fashion, the organized configuration in space and time of physiological processes, integrated to a high degree, takes on a new form or character that is no additional entity, no added substance or process, but just the *form* of coordination and integration of the physiological activity by which the organism maintains its integrity. And this new form is sentience or feeling. Perhaps it could as well be called a distinct "state" of the system, as gaseous, liquid, and solid are distinct states of chemical substances.

Hegel identified sentience (*Empfindung*) with the soul, and Aristotle maintained that the soul was the form of the body. This, it seems to me, is sound doctrine, and one that I propose here to repeat. The soul (if we are to retain the term and the concept at all) is not a separate "thing" (what Hegel sarcastically castigated as *ein Seelending*) attached to, or associated with, the body, acting upon it from the outside, or acted upon by it to generate sensation. It is the form of the body, the new quality evinced at a specific, critical threshold of intensity of integrated physiological activity. Feeling, the self-revelation of this new form, is not just something triggered in particulate flashes by special processes in the nervous system; it is basically bodily feeling, the body as felt (what Merleau-Ponty calls "the lived body," though here its "lived" quality is more than mere metabolism or physiology, but is sentience).

THE SENTIENT WHOLE

Sentience is not only the feeling of the integrated physiological whole of the body but (inasmuch as the organism is auturgically self-maintaining and so responsive to the physico-chemical, as well as the organic symbiotic, influences impinging upon it) is also the feeling of all these focused into a single complex whole. In its primitive form, sentience is one, single and indivisible whole of feeling; and, as it includes both the feeling of the body and of all the organic and inorganic effects upon it, of the environment, it is the same whole as we have been contemplating throughout this essay, now raised to a higher degree of unity and integrity. The physical whole is reflected and registered in a more autonomous form in the biosphere, and the biosphere is again reflected, along with the physical surround, in the organism, and this totality is brought to the level of sensation in bodily feeling.

Such registration of the natural world in sentience is copiously exemplified (though at a higher level) in the felt response which we ourselves experience to seasonal changes, the weather, and climatic conditions. The alternation of night and day and the accompanying rhythms of sleeping and waking are further testimony to the registration by the body of external conditions and its felt response to them. Moods change with the weather and the seasons; energy rises and falls with the earth's diurnal rotation, and is felt accordingly. We see and can empathize with similar responses in other species of animal, which hibernate, migrate, or mate according to the season of the year. All this is obviously inseparably related to the flow of energy into and through the organism from external physical sources, and to the felt needs of its body and their supply through its physiological and behavioural activities.

PANPSYCHISM, DUALISM, AND IDENTITY

Initially, the sentient whole is one indiscriminate conglomeration of feeling tones, felt as a single mood or "temper," the fused combination of the varied tones and modalities as yet undistinguished and subconscious. Because in primitive sentience, as such, there is no discrimination of feeling, self and other (subject and object) are not mutually distinguished, so there is no intentionality. Accordingly, primitive sentience must be pre-conscious, but it is the material content of all consciousness and becomes its immediate object.

How far down the evolutionary scale sentience occurs, and at what level consciousness proper emerges, is of necessity a matter of speculation and can only be inferred from the behaviour of the organic body. It is hard to believe that the behaviour of *Paramoecium*, or of *Stentor*, or of *Arcella* is not prompted by sensibility to outside influences. How (we may ask) is the response to lack of oxygen possible unless it is somehow sensed? No inorganic reducing substance can migrate to seek an oxygen supply, however we may imagine that it is in some way sensitive to the presence of oxygen when that is available.

Such imputation of sensitivity in physical bodies to physical forces (electromagnetic in chemical attraction) is only metaphorically justifiable, except for those who espouse panpsychism, which many eminent thinkers have done in some form or other (Spinoza, Leibniz, Schelling, Whitehead, Teilhard, and more recently, Timothy Sprigge, for example), but the hypothesis is not necessary if one regards holism as a matter of degree. I am here presenting sentience as a higher degree of integration than mere physiological or metabolic cycles, and these again as marking a higher degree of wholeness than simple chemical combination or physical cohesion. At lesser degrees of integrity other qualities prior to, and possibly, in various ways prefiguring sentience reveal themselves, but sentience has to await a special and specific degree of wholeness before it emerges.

What this conception may have in common with panpsychism is its insistence on continuity between the graded phases of self-specification of the organizing principle, which is universal to the entire scale. There can be no question that the relation between sentience and consciousness, at any rate, is one of degree, if only of clarity and articulation, and that in the evolutionary scale the latter must have emerged out of the former gradually, and probably concomitantly with the development of brain capacity and organization.

This doctrine has the advantage of disposing once and for all of the problems attendant upon body-mind dualism, which may still haunt even panpsychism. However prevalent sentience might be, the relation between the physical and psychic remains problematical, but if sentience is regarded as no more (nor less) than the form of the neuro-physiologically whole at a critical intensity of integration, questions about interaction, parallelism, or epiphenomenalism fall away or are immediately answered. If sentience is a higher degree of unity than physiological organicism, it will transform physiology into behaviour, which is a single activity of discriminative response to a presented situation, not an interaction between neuro-physiological process and psychical experience. Psychical experience may, if one so wishes, be thought of as an epiphenomenon, but not as an epiphenomenon that makes no difference to the process to which it gives phenomenological status; nor are there two separate entities, a body and soul (or mind), that have to be related to each other, whether as acting in parallel or by mutual influence. There is only one.

On the other hand, what I am advocating is not to be confused with the so-called Neural Identity theory. That, it seems to me, plays fast and loose with the sense of the word identity. It is alleged that there is only one reality but that it is describable, and is described, in two different languages; and that is palpably false. To describe the sensation I experience when a dentist exposes an unanaesthetized nerve is quite definitely not to describe the neural reaction he causes. That, the neurophysiologist assures us, when measured by means of an electrode, is barely noticeable, whereas my sensation is one of extreme and excruciating pain. In like manner, when I see and describe a circle drawn on a blackboard, what I see and describe is circular. But the neurophysiologist discovers that the excited area of the brain is divided into two parts, one in each hemisphere, and that neither part is circular nor even semicircular, but that each is an elongated horseshoe shape. Moreover, whereas the circle perceived lies in a plane perpendicular to the line of sight, the affected cerebral areas are aligned parallel to the line of sight, on either side of it, with the open ends of the horseshoes towards the rear. What now is supposed to be identical with what? In what sense are they identical, since neither has properties or qualities like the other? Furthermore, neurophysiologists can find no difference in character, but only, at most, a difference in disposition and pattern between the nerve impulses accompanying sensations of sight and those of hearing, nor between either of these and those of smell or taste. Yet the sensations are so different that comparison between them defeats all efforts to express it in words. Again, in what sense of identity can it be said that the sensations are identical with the neural activity, or that the specifically different linguistic expressions describing them are describing an identical reality?

The view I propose is that there is indeed only one reality but that it displays itself in a series of forms with different degrees of unity and wholeness. At each successive level, the entity (or entities) concerned display different qualities and capacities, although they presuppose and involve all the prior forms and degrees of actualization. When we reach the level of mind, these qualities are sensory, as at every prior level they are not. It seems to have been something of this sort that Galileo, Descartes, and Locke foreshadowed when they distinguished primary from secondary qualities. But whereas earlier philosophers tended to stress what they called the mind-dependence of secondary qualities, as if they were produced by the activity of a separate agency from the physical, I prefer to hold that the difference is due to the same activity that has been operative throughout, active now with a higher degree of unified complexity and self-determination. At all the lower stages, the qualities are implicit and latent, but at that of mentality they become apparent and self-manifesting. Accordingly, there is a duality of degree in intensity of integration between the exclusively physiological and the psychological, and a corresponding duality of qualitative form, but there is no dualism of substantive existence.

PRIMITIVE SENTIENCE

In primitive sentience, then, the organism has internalized the diversity of the physical and biotic world as a single confused whole of feeling that is pre-conscious and pre-perceptual. In it are merged undistinguished all the various sense modalities, somatic, visceral, kinaesthetic, proximal, and distal, to form what may be called the psychical field. This should not be conceived (although it may legitimately be imagined in simile) as a kind of glow illuminating the felt content of organic activity. It is a continuation, in a superior phase of integration, of that activity itself. It is the activity of organization at a higher pitch of intensity and focused in a greater degree of centreity.

I am especially concerned to stress the continuity of the dynamic principle and its *energeia*, its organizing activity, operating at successive levels of self-specification that, as they supervene one over another, express the universal nature of the principle more and more adequately; so that what emerges at the level of mentality is the same principle of organization as has been operative at the physical and biological levels and is now becoming aware of itself and its own spontaneous activity. Its self-specification is at every stage structured as a scale of forms, so that the major scale discloses itself as a scale of scales, a system of systems, a whole of wholes mutually in dialectical relation—the physical spatio-temporal field, the biotic morphogenetic field, and now the psychical field.

Primitive sentience cannot be psychologically isolated, nor can it be described in its purity, just because it is an indiscriminate amalgam of feeling, whereas awareness requires discrimination, and description involves distinction of qualities. That there is, however, such a primary phase of sense-feeling is evidenced by the unfailing accompaniment of every kind and level of awareness by emotional tone and sensory content. The psychologist, when seeking to reduce the sensory field to homogeneity, is attempting to discover sentience in its primitive phase.[5] Such attempts have invariably failed, but some of them have revealed sensation approximating to a uniform field, and do give evidence of a pre-perceptual level into which attention introduces some degree of articulation.

Moreover, psychologists have long insisted that every mental state involves three inseparable (if distinguishable) aspects: emotive, conative, and cognitive. The cognitive, as we shall presently find, involves the selective, differentiating, and ordering activity of attention, but the other two invariable accompaniments are the persistence, and in some special cases the residual echo, so to speak, of primitive sentience.

Experimental investigation indicates that the mid-brain (the hypothalamus, the thalamus, the amygdala and hippocampus, and the reticular formation of the brain stem) is the seat of the neural integration of impulses coming from all parts of the body. These areas of the brain apparently together form the immediate bodily counterpart of sentience,[6] supporting the suggestion just made that sentience is rather an activity of organization than some sort of diffuse illumination of an already existing content. There are also indications from neurophysiological investigations that the activity in the reticular formation becomes, as it were, self-enfolded or peculiarly compexified, so as to enhance or inhibit afferent and efferent impulses. Thus, it may well (in conjunction with cortical activity) be the physiological mechanism corresponding to the concentration of attention,[7] which, as psychically experienced, proves to be the selective and distinguishing agency effecting a further stage of organization and initiating the next level of mental activity. The transmission, integration, and self-enfolding of neural processes in these structures of the brain discriminates between influences impinging upon, and impulses internal to, the body, and their functioning is revealed psychologically in the direction of attention by instinctive urges, such as those that register seasonal changes and environmental needs.

THE UNCONSCIOUS

Primitive sentience, being indiscriminate, is necessarily pre-perceptual, which means that it is precognitive. Does this mean that it comprises what has been called "the unconscious mind"? What is meant by that phrase is, however, unclear. There seem to be at least two levels of unconscious mental activity: one of which there is no direct cognition whatsoever, the storage, for example, of long-term memory and the process of its recall when required; another of which we are dimly aware (or can become so), as when some sensation or idea is held at a very low level of attention (e.g., the sound of the sea on the seashore, to which we become accustomed and no longer hear unless attention is redirected to it). Much mental activity does seem to go on at the lower level, and sometimes activity of a highly sophisticated kind, as occurs in rare cases of prodigious calculation feats performed without the conscious participation of the subject. Many eminent thinkers and artists have given testimony to the experience of finding the solution of difficult problems, or the perfect accomplishment of a work of art in a flash of inspiration at the most unexpected and inappropriate moments, after long and sometimes apparently fruitless conscious efforts. The discovery had evidently been accomplished unconsciously, without the conscious participation of its author.

How such unconscious activities might contribute to the sentient field is far from clear, but the account that I have assayed to give of primitive sentience does not exclude the possibility of their doing so, for they almost certainly do involve complicated brain processes, which perhaps become fused into the total neurophysiological complex only at a higher degree of integration, when they rise to the level of cognition. Perhaps some explanation along these lines might be given of experiences like that of blindsight, when damage to a portion of the visual cortex causes blindness in part of the visual field, although the subject can become inexplicably aware of the general nature of an object presented in the area of blindness. This is apparently because of the registration in some way of the image projected on the retina of the eye in other parts of the brain, such as the lower temporal lobe. What is unconsciously registered somehow becomes available to consciousness.

ATTENTION AND CONSCIOUSNESS

Because sentience has a diverse and variegated content that (as merely sentient) is indiscriminate, it is in contradiction with itself. On the one hand, its unity demands the amalgamation of its constituent qualities into a whole, while, on the other, its wholeness implies and requires differentiation. The contradiction needs to be resolved, and the sentient organism fulfils this need by turning its sentient activity back upon itself. The felt impulse of the organism to maintain its systematic coherence, differentiated as it is into specific urges, directs the sentient activity upon special features that answer to its present demands, so that it concentrates itself in the selective operation of attention.

Attention selects an element within the felt whole, distinguishes it from the felt background, and creates a figure-and-ground structure within the psychical field, making it an object for consciousness, which is in this way directed upon it. Nothing less than a figure against a contrasted background can be an object for consciousness, and consciousness is always a matter of degree. It dissipates and is submerged in the subconscious if this minimum of structure is lost. Consciousness thus varies in clarity and definition with the degree of sharpness and articulation of attention.[8]

There is no consciousness without an object upon which it is intentionally directed. Whereas sentience does not, consciousness does imply the distinction between subject and object. The object is, as it were, projected and held "before" the subject, which contemplates it as a whole.

Consciousness has been compared to a searchlight playing upon successive objects and illuminating the surrounding landscape. For many purposes this is an apt and felicitous simile, both for attention and for awareness; but it can also be misleading, because, as a simile drawn from vision, it attributes to consciousness as a whole the character of one prerequisite (light) of one special modality. The congenitally blind are as fully conscious as the sighted, although, unlike them, they do not habitually visualize the objects of the other senses. Vision is not indispensable to consciousness, even though it tends to predominate and is paramount among the senses, so that those endowed with it interpret other sense impressions in terms of sight—visualizing the unseen objects that are heard or smelt and the shapes of things felt in the dark.

Like sentience, on which attention plays, consciousness is an *activity*, but now in the succeeding phase, which is one of analysis and synthesis. It is what H. H. Joachim perspicaciously called an analytic-synthetic discursus.[9] It is no mere passive acceptation of information offered from an external source; it is not simply a "gaze" or "regard" (as Husserl so frequently represents it) playing like a lightbeam on its object. It is an activity of distinguishing, relating, and ordering elements in the psychical field. It is, once again, the activity of organizing. In essence it is judgement, the primitive form of which is "This, not that"—an identification coupled with a distinction. This kind of activity is common to all the sense modalities as consciously discerned, all observation, and all thought.

The agency and instrument of organization, attention, introduces structure into the sentient field by focusing upon an object, more or less acutely, and relegating the rest of the field to successively lower levels of prominence. For instance, when one is reading a book, one concentrates on the meaning of the text, while the actual shape and order of the printed letters occupy a lower level of attention. Distant sounds of traffic or of children playing are heard as mere background noise. Vaguer still are the sensations of pressure from the chair on which one sits, and still less is one conscious of feeling clothes on the body. What is in focus may, according to circumstances and the degree of interest, be more or less prominent, and the lower levels of attention may be similarly more or less depressed.

Present consciousness always has this hierarchical or pyramidal structure, yet is also capable, as it were, of extricating itself from it in some mysterious way and of grasping the entire structure as a whole, as it must in order to envisage its general form. In this way, as we shall repeatedly find, consciousness is self-transcendent, at the same time immediately aware of some object or complex, and concurrently apprehending (or being capable of doing so) its own nature and relation to that object.

Attention, creating the object, by singling it out of the psychical field (before which selection and distinction it could be no object), is thus the initiation of consciousness. To revert to the simile of light, we might say that it is attention that turns on the switch. It does so, in the first instance, in the service of the organic urge and need of the living body, but then, as mental activity develops, in response to perceptual demands (as when the infantile gaze follows a moving object, or when the child first distinguishes and seeks to manipulate what it discerns before it), and subsequently more and more in the service of intellectual investigation.

COGNITION

Only after a minimal figure-and-ground complex has been singled out by attention from the sentient field can the experience be described as cognitive. Here perception is born, the inchoate subject confronts an object, and a beginning is made of the organization of the psychical whole into an experienced world. The articulation of the sentient field is accomplished psychologically through a series of overlapping stages, in which objects are distinguished, among them the body itself as the seat of the sensations from which they are constituted. Concurrently, the various sense modalities (as differing sense fields) are distinguished. As objects are related to one another and to the body in which sensations are felt, self becomes opposed to not-self, and an outer world is built up in which the subject is conceived as one member and the organism that it inhabits is placed in relation to its encompassing environment.

So arises the perceptual world of common sense—what Husserl and his followers call "the life-world," constituting the perennial background or "horizon" of all perception. In "the natural attitude" (to retain Husserlian terminology), it is accepted as "given" and is observed as external to the consciousness in which it is revealed. At this stage, its structures and contrasts are simply taken for granted, as they immediately appear, and it is only when contradictions arise in the common-sense acceptance of these relationships that a new stage of reflection is reached, at which questions are raised, explanations sought, and the mind embarks upon theoretical inter-pretation.

It is in virtue of the self-transcendent character of consciousness that the mind reaches the point at which a fresh transition, a further self-enfoldment, takes place: the stage at which self-reflection is achieved. This is an accomplishment to which only humankind, among the animal species, has so far attained (although gorillas and chimpanzees show some sign of having reached its onset). It is the crucial point at which the self becomes aware of its own identity and knows itself as "I," at which it makes itself, along with its ideal content, object to itself as subject. Here the mind enters upon the stage of self-reflection: reflection upon the nature of its objects and its own relation to them. This is the dawn of intellect, the birth of wonder, and the awakening of self-criticism and self-appraisal.

Reflection is the distinguishing mark of the human. Without it there can be no morality, no civil society, no science or philosophy, no art or religion. The answer to materialists, behaviourists, sceptics, deconstruc-tionists, and any who seek to deny the validity of presupposing the existence of a subject in knowing, is that without self-reflection there could be no materialism, no behaviourism, no scepticism, and no theoretical deconstruction; and the first fruit of reflection is the indefeasible revelation to the self of its own existence. To this critical development we shall have to return, for it is crucial to the anthropic principle and its philosophical implications.

PRELIMINARY MISGIVINGS

Those who remember the traditional idealistic problematic will no doubt wish to challenge this account of the emergence of knowledge, to ask how, if the life-world is thus constructed from the contents of the sentient field, we can ever know whether anything in the outer world corresponds to our subjective construction. The question, however, is misplaced and mis-guided. Objective and subjective is a distinction made within the life-world, which the experience embraces as a whole. As Husserl and others before him have insisted, we can in no way get outside our own consciousness.

There is no outside, if only because outer and inner is an opposition constituted within experience.

The life-world is all-inclusive, and while in the natural attitude the world, as perceived by common sense, is regarded as "external" to the mind, that is because, at that level, "the mind" is imagined as a function of the brain and is objectified along with the body. The subsequent attempt to explain consciousness as the result of the transmission from external objects of physical impulses through the senses to the brain proves incoherent and has brought (in the history of philosophy) only epistemological disaster. What has been overlooked is the self-transcendent character of consciousness, aware at once of the presented object and of its own relation to it. Thus, as it distinguishes subject from object, it also grasps their relation within the whole, which together they constitute. The mind, become self-conscious, is capable of developing the implications of such holism in philosophical reflection. Any account given here of the nature and conclusions from such comprehension must necessarily be proleptic, but to allay misgivings, some brief anticipation may be offered.

The world disclosed in observation and interpreted in science and philosophy reveals itself as a dialectical scale of forms, primarily in experience, ranging from sentience through perception and reflection to comprehension. From this it transpires that the later phases are the more adequate expression of the whole than the earlier, and reflect more truly the real nature of the whole. Consequently, the scientific interpretation of common-sense deliverances (or observation), to go no farther, is the truer representation of the reality than the common-sense view. Science, as we have discovered, presents the world as a whole, specified in a scale of forms, that ranges from the physical to the organic and sublates all of these in sentience, so that the sentient field is the same whole, in a higher phase of self-determination, than those that have preceded. This is why we cannot get outside the consciousness that arises from primitive sentience, and why the life-world is an all-inclusive whole. It is because the physical world is an all-inclusive whole ("finite but unbounded") outside of which there is nothing; and the experienced world is that same whole become aware of itself. What "corresponds" to it, therefore, are simply the prior phases of its own development. These go back beyond sentience for the very reason that sentience has revealed itself as the form of the body, the reflection and registration of organismic activity, integral to the biosphere and rooted in a physico-chemical environment. The object of the mind is, therefore, its own self in becoming, and the subject is no less than the world come to consciousness of itself. Subject and object are identical, and fact corresponds to theory just so far as the theory is what the fact itself has become in bringing itself to consciousness. This conclusion reveals itself in reflection upon science and experience in general at the philosophical stage.

In thus bringing itself to awareness, the universal principle of wholeness remains immanent, as subject, in the experience; and without such immanence the experience could not be one. The immanent universal is what Kant called the transcendental synthetic unity of apperception. It is what constitutes the self, as distinguished within the sentient and conscious whole, a transcendental *ego*, transcendentally aware both of itself and its other, and cognizant of the whole immanent in its own experience of the world.

Dogmatic idealism, which fails to recognize the dialectical character of the whole immanent in finite experience, commits the epistemologist to a subjectivism that is as disastrous as dogmatic realism: the first because it leads inevitably to self-contradiction in solipsism; the second because, by confining consciousness to an effect in the brain of an assumed (but *ex hypothesi* unknowable) external cause, it excludes from knowledge the very object that knowledge seeks and claims to embrace. The one feasible resolution of the contradictions involved is in the self-specification of the universal whole as a dialectical scale of forms, manifest in the physical universe and bringing itself to fruition through the organic world in the self-consciousness of intelligent life.

What, then, is to be taken as the criterion of truth? By what standard do we assess the validity of our knowledge of the world? It is the degree of coherent wholeness of the experience judged, both observation and theory together. When they do not agree, contradiction arises, due to some oversight or omission in one or the other, and corrections are needed, or presuppositions must be changed, in order to restore coherence and systematic wholeness.[10] But to make pronouncements of this kind is to anticipate what can only properly be justified at a later stage and is inserted here simply to assuage the premature misgivings of possible objectors.

ARTIFICIAL INTELLIGENCE AND CONSCIOUSNESS

To speak of mentality in terms of sentience and consciousness is nowadays anathema to many. A long history of materialism, mechanism, and behaviourism (largely a hangover from the Renaissance world-view) has resulted, more recently, in an enthusiasm for artificial intelligence and the opinion that the human brain is some kind of highly complex digital computer (or general Turing machine), to the functioning of which consciousness, if it exists at all, is irrelevant.

To doubt or deny the existence of consciousness (we have known since Augustine and Descartes) is self-refuting. It is that of whose existence we are directly assured by its very occurrence, and without which we could be assured of nothing. No theory of artificial intelligence, no opinion about the epiphenominal character of awareness, could be entertained without it. Moreover, the behaviourist, demanding cognizance only of what can be "publicly observed," disregards the fact that such observation, so far as it is perception (as it must be), is the private experience of the observers, and that they can communicate it only on the assumption that others can become aware of their means of communication. All this, again, presumes the use of the senses and, therefore, the existence and presence of sentient experience.

These facts, however, do not disconcert the advocate of artificial intelligence (AI), which in every one of the computers operating ubiquitously today makes no claim (at any rate thus far) to consciousness.

The tradition of behaviourism has prompted what has become known as the Turing test for artificial intelligence. How do we ever become assured of the presence of consciousness in any but our own selves? (Observe that we must confess to being assured of our own awareness.) Only by what it is supposed to do in the external observable world. If, then, the answer to questions given by a hidden computer could not be distinguished from those of a professedly conscious agent, how would the two differ?

What this type of reasoning seems consistently to overlook is that no computer can answer any question unless it has been appropriately programmed; and that requires the encoding of the input into the requisite computer language. No answer given to a question by a computer can be understood unless its output can be decoded and the decoded message can be read. And no computer has yet been, or presumably could be, devised that can encode its own input and decode its own output. Even if one could be constructed to do all this, it could be so only by a human intelligence with conscious intent and capability. To presume that the human brain is after all only a sort of computer would then be to beg the question and to argue in a circle (unless one were prepared to entertain the belief in some superhuman creator who was not a machine).

A general Turing machine, the theoretical archetype of all computers, does no more than operate a mathematical algorithm; that is, a procedure in accordance with set rules. It may be complex and sophisticated to any desired degree, but it must be recognized as appropriate and valid for the purpose (or purposes) for which it is used, and such recognition cannot be the result of any such procedure. Moreover, no algorithm (for similar reasons) can be devised except by a human mind, the ingenuity of which is always tacitly presupposed by the existence of a Turing machine: no Turing machine, we may confidently assert, without Turing. So, once more, if we try to pretend that the human brain is no more than a complicated computer, we beg the question.[11]

In his book *The Emperor's New Mind*, Roger Penrose has convincingly demonstrated that there are certain aspects of mathematical thinking that are not computable—that cannot be the result of algorithmic calculation. Most significantly, Gödel's Theorem proves that in any formal system whatsoever, a legitimate proposition can always be formulated that is unprovable in the system, and so it establishes that there is some mathematical thinking that is not formalizable and therefore cannot be computable. The Gödel proposition itself states that it is not provable within the system in which it has been formulated, and this at once is seen to be true. Seeing it as true requires insight, which, since the proposition cannot be formally proved, is not formalizable and not computable. Our insight is not the product of this or any other algorithm. Thus it becomes obvious that the human brain cannot simply, in all its functioning, be a computer, even if other considerations lead us to conclude that in some of its unconscious operations it may act like one.

Penrose produces numerous other examples of mathematical thinking which defy computerization, and he makes a powerful case for the necessity of consciousness in human thought wherever insight is required. "To understand fully why an algorithmic procedure does what it is supposed to involves *insights*," he writes, emphasizing that word; and he asks whether insights are themselves algorithmic.[12] Gödel's theorem both proves and exemplifies the general fact that they are not. Moreover, Penrose points out, it is not sufficient for an algorithm to be operated by a machine. If it is to be significant and useful it must be interpreted, which again requires non-computable insight.

That insight involves consciousness must be accepted if we recognize consciousness as the activity of organizing sentient presentation. In the first place, such organization is the establishment and comprehension of relationships within a whole, and the perception of relations precisely is what constitutes insight. In the second place, the relations are established between elements in the sentient field, and sentience is what characterizes all consciousness. Thinking, including mathematics, continues this activity at a high level of abstraction, but is never wholly devoid of sentient content.

But how, we may be asked, have we established the existence of sentience itself? To this the answer is: By the self-certainty of consciousness, the presence of which is undeniable without self-refutation (for one must be conscious to deny it, and could not postulate it without having it). And consciousness is nothing other than the awareness of elements in the sentient field. Of "raw feelings," while we have them, we cannot deny the existence, and if we did not have them, there would be no possibility of presuming or of inventing their existence. Methods of investigation that ignore the occurrence of sentience and consciousness, or which refuse to make reference to it, may in some circumstances, and for acceptable reasons, be justified, but the pretence that sentience and consciousness do not exist can only be an affectation on the part of those who seek to deny what, by its very nature, is ineluctably manifest to themselves and to all other cognizant beings.

Notes

1. Cf. F. W. J. Schelling, *Ideen zu einer Philosophie der Natur*, translated as *Ideas for a Philosophy of Nature* by Errol E. Harris and Peter Heath (Cambridge University Press, Cambridge and New York, 1988), pp. 30–34.
2. Cf. Ch. 2, above, pp. 25–26.
3. Cf. Susanne Langer, *Philosophical Sketches* (Johns Hopkins University Press, Baltimore, 1962), p. 9; and *Mind: An Essay on Human Feeling*, Vol. I (1967), Ch. I.
4. Cf. B. Blanshard, "Harris on Internal Relations," in *Dialectic and Contemporary Science*, ed. P. Grier (University Press of America, Lanham, MD, 1989).
5. Cf. K. Koffka, *Principles of Gestalt Psychology* (Routledge and Kegan Paul, London, 1955), pp. 323–325; W. Metzger, *Optische Untersuchungen am Ganzfeld, Psychologische Forschung*, 15, 1930.
6. Cf. *Reticular Formation of the Brain*, Henry Ford Hospital International Symposium, 1958, ed. H. H. Jasper et al.; H. E. Himwich, "Emotional Aspects of Mind: Clinical and Neurological Analyses," in *Theories of the Mind*, ed. Jordan Scher (Free Press of Glencoe, Chicago, 1962).
7. Cf. H. H. Jasper, "Ascending Activities of the Reticular System," in *Reticular Formation of the Brain*, Ch. 15.
8. Concentration of attention, and the consequent degree of consciousness, may be affected in its turn by bodily conditions, including the effects of drugs, shock, or disease.
9. Cf. H. H. Joachim, *Logical Studies* (Oxford University Press, Oxford, 1942).
10. See Ch. 9, p. 132ff, below.
11. Cf. my "Minds and Mechanical Models," in *Theories of the Mind*, ed. Jordan Scher.
12. Roger Penrose, *The Emperor's New Mind*, p. 42.

9
Observation

MISCONCEPTIONS

Primitive sentience may be compared to a *camera obscura*, in which the entire external surround is reflected through a single lens on to a screen within a limited space. The simile, like so many others, has its uses and its dangers. It is helpful as an illustration of the way in which the entire universe is, as it were, focused and concentrated within the individual organism. But it is also misleading: first, because sentience is an indiscriminate mass of diverse feelings, whereas the image in the *camera obscura* is that of an articulated scene; and second, because of an even more hazardous discrepancy for epistemology, the omission from the image in the camera of the viewer, who sees, recognizes, and interprets the reflected objects. It thus lures us into the misconception that perceiving is the result of the transmission of physical effects from the outside world, through our sense-organs, to create some kind of replica or model in the brain.[1] This, as has already been noted, is a theory unsupportable even by neurophysiological evidence. Epistemologically it is totally disastrous, with implications that are utterly incoherent.

To discern the true nature of perception, we may do well to begin by determining what it is not, and why. In the first place, perception is not a simple end result of a physical and physiological causal chain, as Locke assumes. "If then," he writes, "external objects be not united to our minds when they produce ideas therein; and yet we perceive . . . such of them as fall under our senses, it is evident that some motion must be thence continued by our nerves, or animal spirits, by some parts of our bodies, to the brains or seat of sensation, there to produce in our minds the ideas we have of them."[2]

Quite apart from the difficulty of explaining how the end result of a purely physical process can be converted into a psychical cognition, no such description can be a true account of perception, because the percept is assumed to be the end result of a causal chain, and as the causal relation between the object and the percept is excluded from this end result, it could never come to be known, unless some other means of observation than perception were available to us (which Locke and others who hold similar views deny). The percipient is certainly never aware of any such causation, and no observations of what occurs in other persons, concomitantly with their reports of what they perceive, can ever be anything other than percepts, the causes of which are equally beyond the reach of the percipient. Since any percept in itself can as well be delusory as veridical, the truth of this account of the matter could never be ascertained. If it were true it could not be known, and if it is known it cannot be true.

In the second place, perception cannot be what the above-mentioned causal theory usually requires as its complement; that is, that the product of the causal process, the "idea," should be a copy or representation of the external thing taken to be its object. To recognize an image as a copy of some archetype, we should need access to the archetype other than that provided by the representation itself, in order to be able to make the necessary comparison. And no such independent access is available to us. If we could know that the idea was a copy of its object, we should already know what the object was like and should need no idea to apprise us of it; but, as the idea is all we have, the copy theory is untenable.[3]

In the third place, it is usually assumed, as a corollary of this type of theory, that the sensation caused from without is an indubitable datum. But perception cannot be such immediate acceptance of data, assumed to be indubitable (or "hard"). All that is indubitable about any experience is that it occurs when it does. If I experience a green flash, I cannot doubt while I am having the experience that I am having that type of sensation. But I cannot be sure that I am "seeing" a green flash if that means that I am perceiving something occurring independently of my sensation. The sensation may be no more than the mental accompaniment of a neural discharge in my occipital lobe of which directly I am totally unaware, and of which the cause may be entirely different from and unconnected with any light flash (e.g., the chemical effect of some drug). What is indubitable in sensation, therefore, is no hard datum from which veridical perception might be constructed.

The immediate sensum, moreover, is not and cannot be apprehended as such, unless it is distinguished from a background and identified as an object, an accomplishment requiring inchoate comparative judgement. Awareness of it, then, cannot be mere acceptance of a "given" but must involve some sort of discursive activity—some degree of rudimentary thought.

SENSE-DATA

Theories premised on the belief in "hard," sense-given data as the building blocks from which percepts are constructed are generally known as sense-datum theories. Their error lies not in the contention that we experience sensa, like colour patches, tones, pains, prickles, and whiffs of perfume, but in the belief that these are simply given to and accepted by the mind as the unrevisable bases of all knowledge. That such sensa do occur and are universally experienced cannot and need not be disputed. No more eloquent plea for their existence has been made than H. H. Price's exemplary paper, "Appearing and Appearances."[4]

The ubiquitous experience of sensa is the probable reason for the persistence of sense-datum theories of perception, despite the cogent criticisms that have been levelled against them ever since Kant demonstrated that the manifold (of sense) contained in every intuition can be represented even as a manifold only by an act of *a priori* synthesis.[5] Such criticisms have been voiced mainly by idealists, but other objections have been raised more recently from rather different quarters. These have often been misplaced and can be more easily answered.

The problem which perception has posed for philosophers ever since the Greeks, and (in the East) even earlier, has been how knowledge of physical objects external to the human body could be acquired through the senses (or how representations of them could get into the mind). When we perceive such objects, we perceive them whole, and we know within reasonable limits what they are like; yet what the senses immediately apprehend are but colours, shapes, sounds, scents, tastes, and the like, none of which by itself exhausts even the apparent nature of a physical object. Several such sensa taken together are still insufficient to do more than indicate what the object intrinsically is, and most of them are quite unlike what we come to know as the actual qualities of the thing observed. Yet, when we perceive physical things, we seem to grasp at once what it is with which we are confronted. It is in the attempt to give some coherent account of the relation of qualities directly sensed to the actual objects perceived that some philosophers have devised sense-datum theories.

Contemporary critics have complained that sense-data are queer "objects," of which people are normally unaware and to which nobody can point, by way of explaining what is meant by the phrase, as one might point to the fovea in a dissected eye to explain what is meant by that word.[6] Those critics also allege that when a physical object appears to us perspectivally we can intelligibly communicate the fact simply by saying that we see (for instance) a round plate, but that it looks elliptical from where we are observing it, and that there is no need to invoke sense-data to explain the appearance. All this is no doubt true, but it does not, as the critics often claim (or at least imply), remove the original problems.

Such critics commonly use the word "see" to imply that what is seen is an external object that is either perceived veridically, or "seen as" what we take it to be. The verb is, however, ambiguous in common use. Although very generally used in one or both of these senses, it is also often used in describing the experience of illusions and hallucinations, as when we say that a blow on the head makes us "see stars," or when somebody claims to have "seen" a ghost (quite apart from "seeing" after-images). This ambiguity tends to bedevil the entire debate about sense-data.

Now, the main problem of perception is to determine the criterion by which we distinguish this kind of seeing from veridical seeing (using the word as the critics of sense-data prefer); and it does not help in the least to say that in the case of illusions or hallucinations we only "think we see" the spectral appearance. The subject of the experience will protest that he or she has indeed seen what appeared, and must be convinced by some other means that it was not real. We need some valid criterion to judge between cases in which we actually see (veridically) and those in which we only think we see. It was to cope with this kind of problem that the sense-datum theories were devised (witness their common use of the so-called argument from illusion). But the critics almost invariably ignore this central problem.

In an elegant paper, Professor Winston H. Barnes has found fault with the sense-datum idea, for very good reasons, and has expressed a preference for what he calls (along with other such critics) the language of appearance— admitting, however, that it does not resolve all the problems of perceiving.[7] But Professor Roderick Chisholm has shown that the language (and the theory) of appearing give rise to similar, and additional, metaphysical problems to the language and the theory of sense-data.[8] Many other criticisms follow similar lines but are all inconclusive and unsuccessful, in part because we undoubtedly do experience sensuous phenomena that may, not infelicitously, be termed sensa, and in part because the sense-data theories draw attention to the very important fact that in perceiving we become aware of far more (e.g., that I now see a round plate which, from this angle, looks elliptical) than we can possibly sense immediately in any single apprehension, and to the equally important problem of how what one can directly sense is related to what one thereby comes to know in the actual perception. For instance, how do I *know* that what looks elliptical from here is actually round? Those acquainted with the Ames demonstrations, in which perspectival distortions are used to create seemingly veridical appearances, will appreciate the importance of this question.[9]

Perhaps the best known and most sustained attack on sense-data theories is that of John Austin, in his posthumously published lectures, *Sense and Sensibilia*. But his arguments are even less considerable than those of the other critics. In fact, this book has so many faults that it ought to be examined in detail, but here that would involve an unwarranted digression. So I shall confine myself to salient points. First, Austin misrepresents the views of those whom he criticizes, apparently deliberately (for he certainly was not ignorant of their writings). For instance, he quotes A. J. Ayer as implying that the "ordinary man" believes he perceives "material things" (such as chairs, tables, pictures, and the like). But, Austin contends, this is not the case, because "the ordinary man" does not believe that what he perceives "is *always* something like furniture, or . . . moderate-sized specimens of dry goods." "We may think, for instance," he continues, "of people, people's voices, rivers, mountains, flames, rainbows, shadows . . ." etc., etc.[10] Indeed, we may, but this is totally irrelevant to the point that Ayer was making, which is clearly stated in the passage quoted by Austin. Ayer's point was that whatever we perceive, the object of which we become aware is not what we directly sense. It is perfectly legitimate for him to use familiar objects as examples, nor is he obliged, in order to make his case, to draw attention to the variety that Austin catalogues.

Again, on page 27 of his book, Austin quotes H. H. Price's *Perception*, where he provisionally defines an illusory sense-datum as one that we tend wrongly to take for the surface of a material object. In this context, Price is admirably precise about what he means and what he is arguing; but Austin cavils that, when two lines of equal length look unequal to us, we do not think or tend to think that we see anything else, and do not normally raise the question whether anything is or is not part of the surface of anything at all. In fact, however, we do at least tend to think that the lines are part of the surface of the page or whatever else they are drawn on, and if they look unequal when they are not, then we do tend to take what we see to be other than what the material object actually is—and that was the point Price was making.

Austin's book is full of such irrelevancies and misrepresentations. To list and examine all of them would be inappropriate and unnecessary to my main purpose, so let these two examples suffice. Austin has committed further, more egregious errors.

On page 12 he argues that a plain man would not consider himself deceived by his senses when confronted with ordinary cases of perspective, with ordinary mirror images, or with dreams. It is not clear how "plain" a man must be before we regard him as utterly thoughtless. Indeed, we have all learned in early childhood to correct, in our perceptual judgements, the effects of perspective, though even in adult life perspectival distortions can produce illusory effects. If we reflect at all, we should certainly admit that perspectival effects are often deceptive. M. C. Escher's engravings exploit such deception in order to create their intriguing effects, and E. H. Gombrich has given numerous other examples.[11] And we seldom accept dreams as veridical but should normally admit at once that they were delusory.

As for mirror images, they always appear to be as far behind the surface of the mirror as the reflected object is in front of it, which is an illusion; and if we trace the outline of the image on the surface of the mirror we find it, surprisingly, to be half the size that we should expect from its normal appearance—so whenever we see mirror images we are deceived by our senses, at all events in some respects. Austin's objections to the argument from illusion (against which he is here contending) must be disingenuous, for they can hardly be attributed either to naivete or to ignorance. Nor is the use of that argument by sense-datum theorists to draw attention to the difference between what is directly sensed and what we take to be actually there illegitimate.

There are other and better reasons for rejecting theories of sense-data than those given by Austin, the most cogent being that, although our senses do deliver to us the kind of sensa that Price and others so tellingly list and describe, they are none of them barely sensed. The least of them are constructed, and many of those alleged by their advocates to be "hard data" are abstractions from more complex objects of more sophisticated perceptual experiences.

That they are at best constructions can be inferred from what neurophysiologists and psychologists tell us contributes to so simple a perception as that of a red patch. The perceived location of the patch in space depends upon the position and movement of the body and on the subconscious relation of the percept to the corresponding proprioceptive and kinaesthetic sensations; it depends on the mutual relation of the images on the two retinae, the accommodation of the eyes and the reflex reactions of the eye muscles, as well as the position of the head. The sensory registration of all these physiological conditions (even though subconscious) is involved in the apprehension of the so-called datum. As Russell Brain writes: "The spatial setting of the red patch is derived from my body; and the proof of this is that when this body machinery goes wrong I no longer see the red patch 'there'. I see two red patches, or the red patch goes round and round, or though I see it I do not know where it is, and cannot find my way to it."[12] So the "sense-datum" is in large measure a product of my physical posture and physiological condition, and it appears as "a red patch there" only to my "normal" vision, the criterion of normality being unspecifiable except by reference to what I now and what I commonly perceive. Thus, it is far from merely given. Moreover, irrespective of subconscious contributions to, and effects upon, the presumed datum, what is perceived is a red (as opposed to a green or a yellow) patch in this light (as distinct from illumination of other brightness or colour), of a definite shape (discriminated from other shapes), contrasted with a vaguer background; there (not elsewhere), here in this room, in Cambridge, or wherever (an unexpressed, yet present, contextual awareness). The percept is implicitly an elaborate judgement, the apprehension of a complex situation that determines the quality, position, and character of the red patch in the focus of attention. Its perceived position varies with its subconsciously estimated size, and its size varies with its distance subconsciously judged by reference to other concurrently perceived objects. The simplest percept is thus a product of quite complicated incipient thinking, as will presently become clear.

PERCEPTUAL KNOWING

This critical interlude has kept us long enough; let us now return to the matter in hand. Sense-data theories are the progeny of Empiricism, which declares all knowledge to be derivative from sense and is then committed to discovering the sensuous data on which it is based. All knowledge is indeed generated from sentience, but what Empiricism overlooks, despite Kant's warning, is that all knowledge is organized experience, and that primitive sentience is below the level of organization essential to cognition, without which there can be no perception. All perception is knowledge and none is mere sensation; it is the awareness that a recognizable object, however simple, is being sensed—that some identifiable object is present to the senses.

There are, in fact, no hard data such as are postulated by the sense-datum theory, because the primal intentional objects are created by that analytic-synthetic activity, which is no less than thought, operating through the agency of attention, as it selects elements in the sentient field and sets them against a background, to which, and within which, it relates them in diverse ways. At the same time, it puts them in relation one to another to constitute a *Gestalt*, in accordance with the principles of organization essential to its nature. Percipient cognition, therefore, is nothing other than a further stage of complexification and intensification of the organizing activity, the progressive forms and phases of which have already been traced at the physical and biological levels. Perception is thus the activity of structuring the contents of primitive sentience; and at every stage, from the most elementary to the most complex, it is always the comprehension of a whole.

Cognition begins with perception, and the *minimum perceptum* is an articulate structure, a figure discriminated against a background, so that even at this minimal level of cognition we begin with a whole. The object is created by attention, which singles out the datum or sensum and distinguishes it from its surrounding field or ground. Henceforth, by successive stages, into the detail of which we need not now enter, objects are identified and distinguished and relations are established between them.

The differentiation of the sentient field takes place in the course of the organism's endeavour to maintain its integrity by adapting to the world in which it lives. Its interaction with that world is registered in sentience, through its physiological processes, and the physico-biological whole into which it has to integrate itself is, as it were, held in suspension within the felt mass of its sentient life. Its movement among and commerce with external things, including other organisms, in its efforts to satisfy its instinctive urges, direct its attention to the relevant elements within the sentient field; and setting them in mutual relation, identifying them, and distinguishing among them is the thinking activity of the conscious subject thus awakened to the apprehension of objects. Such apprehension can take place only through this thinking activity of distinguishing and relating; each individual act being one of judging, initially implicit but, in the more developed phases, explicit and articulate.

The ability to recognize objects, to trace their movements, to adjust motor activity to them, and to manipulate them is a structure of related skills that for the most part have to be learned and are acquired in infancy. Some few may be inherited and instinctive, such as the ability of chicks, immediately after their emergence from the egg, to distinguish edible from inedible grains. But only a few very elementary cognitive abilities are of this kind, and the process of learning to perceive continues late into adult life, resulting among humans in the specialized perceptual skills of the expert, the scientist, and the engineer, who can discern and recognize in technical situations what is entirely lost upon the layman.

The organizing activity depends closely on the fitting of sensory cues into what physiologists and psychologists call schemata. One of the most important of these is "the body schema," especially for perception of the relations between the perceiver and external objects, as well as for the perception of the shapes and positions of bodies in general. This forms the basis for the localization of sensations in the body, the distinction between modalities and that between exteroceptive sensations (on the one hand) and kinaesthetic and somatic sensations (on the other).

Some schemata may be innate, dependent only upon the full development of the sense organs, while others are certainly acquired. Visual depth is probably one of the former, along with the primary spatio-temporal schemata involved in orientation. Whether they are innate or not, and which might be which, is less important, however, than the fact that they are all elements of structure, all wholes. They are in principle what epistemologists call concepts, organizing principles, imposed in perceiving on the sensuous flux by the activity of thinking.

Kant maintained that subsumption of the sensuous manifold under some such concepts was the *a priori* condition of any experience of objects whatsoever. Even the *a priori* forms of intuition, space, and time prove in the later argument of the Transcendental Analytic to be dependent on the concepts of the understanding.[13] And if spatio-temporal schemata are indeed innate, Kant was right to identify space and time as *a priori* forms of intuition and as dependent upon concepts. Today, the experimental work of psychologists bears out Kant's insights, for it reveals that nothing less is ever perceived than *Gestalten*, structured configurations (both spatial and temporal) involving principles of organization. This being so, perceiving and thinking cannot be separated. In Kant's terms, concepts (or concepts projected upon the *a priori* forms of intuition—schemata) are the principles by means of which objects are thought, and so far as such objects are sensuous, the thought is perceptual.

"Intuition," Kant wrote, "without concepts is blind." Again he was right, for sentience is pre-conscious, but Kant ought to have realized that intuition is *never* without concepts. What is "given," if anything, is no more than William James' "blooming, buzzing confusion." The psychologists Borger and Seaborne tell us: "Human sense-organs register light stimuli of varying area, intensity, wave-length and duration, sounds of varing frequency and amplitude and so on. The incoming proximal stimuli are constantly changing, so that from second to second the person receives differing total physical stimuli from the environment. Despite this, perception is of a stable world of solid objects with few, if any, discontinuities."[14] Likewise, E. H. Gombrich: "What we get on the retina, whether we are chickens or human beings, is a welter of light points stimulating sensitive rods and cones that fire their messages into the brain. What we see is a stable world. It takes an effort of the imagination and a fairly complex apparatus to realize the tremendous gulf that exists between the two."[15]

It is the activity of thinking that, by organizing the sentient manifold in accordance with structural principles, generates objects and sets them in relation to produce an orderly world of experience, in the light of which it can then interpret new experiences. Such thinking is always activity, but it is not always (in fact, relatively seldom) explicit and articulate, so it is not commonly recognized as active thought, and percepts are held to be passively received, as a photographic film receives the images projected upon it. But this is a mistake that both psychology and neurophysiology expose and that careful epistemology will avoid.

There is massive experimental evidence to show that the perceived object (whether it be as simple as a colour-patch or as complicated as a landscape) is formed and conditioned by context, spatial and temporal, and by past experience. The activity by which context and experience are made effective is judgement extending into inference, and the evidence is copiously available that these are inchoate in all perception. Idealists have consistently maintained a theory of perception as implicit judgement and inference,[16] but they are by no means alone. A similar view is implied in much that Gilbert Ryle has written on the subject, and in discussions by A. M.

Quinton and D. M. Armstrong. It is also undeniably implicit in everything that Merleau-Ponty so eloquently proclaims about perception.[17] Now, experimental psychologists have produced convincing new empirical evidence of the truth of the position. To give details here would lead me too far from my main theme. I can but refer the reader to the sources in which this evidence is recorded.[18]

The position has, of course, been criticized, in particular by R. J. Hirst, who objects that perception is immediate, without consideration of evidence or deliberation, whereas judgement involves both; that perception is impossible in the absence of its object, whereas judgement is not; and that false judgement is abandoned when the error is discovered, but perceptual illusion persists even when we know it as such. He also submits that we are never aware in perception of judging or inferring and that no introspection reveals these activities.

The first of these objections is invalid against *implicit* judgement and inference, which would not involve deliberation. The second is beside the point because, although judgement is possible in the absence of its object, it does not follow that it is impossible in the presence of its object. False judgement is abandoned when we discover its falsity, because we have also disposed of the misleading evidence, but this is not the case with perceptual illusion, for the distorting features remain before us. More conclusive, however, are the facts revealed by psychological experiments which show that none of Hirst's objections hold good. It is not true that perception occurs without deliberation. In fact, recent psychological theories maintain that it always involves the formation, testing, and confirmation of hypotheses.[19] Bartlett has recorded and described numerous cases in which introspection did reveal thinking, inferring, and deliberation, in efforts to perceive quite simple presented objects.[20] Moreover it is possible with practice to see an illusory structure (like the Muller-Lyer illusion) veridically, despite the persistence of the misleading evidence.

Not only is it false that sense-data are given in perception, but also that the "given" element (as has been maintained by Husserl[21] and at least implied by other writers) is a physical thing. No physical thing is ever presented whole to the senses, yet it is perceived, when at all, as a whole. There are always sides and an inside that are not visible to the eye; shape is not recorded by sound, smell, or taste; even from touch sensations it has largely to be inferred and constructed; and all these sensory cues taken together fail at any one moment to present the whole object as it is perceived. Much has to be, and is, provided by supplementary imagination derived from past experience. As has been said, perception is a skill (or a coordinated group of skills), in large measure learned at an early age, an ongoing activity of judgement and inference that continuously constructs and builds up a coherent experienced world.

It thus seems safe to conclude that perception—as implicit judgement, the cognitive result of an implicit inference issuing from the activity of organization—is the activity of thinking. Another name for this activity is interpretation, that is, the grasp of some item or group of items within the structure of a whole, subject to some principle of order. All perception is interpretive, with reference to the background of the particular percept, both spatial and temporal, that is, in relation to surrounding objects, as well as to past experience; in short, in relation to the funded knowledge of the experienced world.[22] The realization (comparatively recent) by scientists and philosophers of science that all observation is theory-laden is therefore hardly surprising.

THE LIFE-WORLD

The general awareness of the world built up in perception is the common-sense world of "ordinary" experience that has come to be called, following Husserl, the life-world. It is what, in the "natural attitude," we accept as real. It is the experienced world and it must be the object of a unified experience, otherwise it could not be an experience at all. If each presentation were isolated and separate, or even just loosely associated without necessary interconnection, as Hume believed, no experience whatever, let alone of an ordered world, could arise, because every experience, even of the *minimum sensibilium*, is that of a structured whole, which could only be envisaged by a single and unitary subject and could only be cognized as related to other presented objects if both (or all) were held together by the cognizant subject within its own consciousness. For this reason, the experienced life-world is inevitably one world and one experience. It is and must be an integral whole, even if its intrinsic coherence varies in degree according to the extent to which the experience has developed and is systematized by the thinking activity of the subject.

Further, it is the same whole as was encompassed in sentience, but now articulated and raised to a higher degree of organization. Its unity is now not simply organic but is conscious, and is centered in the cognizant subject. In one sense the organization and centreity is imposed upon the experience by the subject, but in another sense it is not. It is not, as it were, brought in from the outside; for the subject is nothing less than the universal principle of wholeness that has been immanent throughout the process of nature, and is intrinsic to the organic unity now come to consciousness through the sentience of the organism. The activity of attention—distinguishing, identifying and interrelating objects—is the same organizing activity as that operative throughout nature, but now no longer acting blindly but consciously, as cognizant thinking. It is the same organizing principle that integrates the physical cosmos and unites the biosphere, which unifies conscious experience and ensures the integrity of the experienced world.

The life-world is thus the same world as the physical world and the biotic world. They are all mutually continuous dialectical phases or specific forms in the necessary self-differentiation of the universal totality. Because the world is a whole it must, of necessity, be complete, both synchronically and diachronically. And as no whole can be complete unless brought to consciousness,[23] the universal principle of structure comes to self-awareness in the consciousness of a cognizant subject, through the natural process that issues in human experience of a perceived world.

The process of bringing the world to consciousness is, in the first instance, the imposition of order and systematic relationships upon the sensory flux. Pre-cognitively, this is not even a flux, and it can be felt as such only as cognized. In the course of becoming organized, self is destinguished from not-self, and the spontaneous activity of thinking becomes aware of its own agency as subject concomitantly with its apprehension of its object, and that object is nothing other than its own self in process of generation. Throughout mental life, the object of awareness is always the prior phase of the dialectical process. In perception it is sentience, articulated into *sensibilia*. This generates the perceptual world of spatio-temporal bodies and their properties; but, as accepted in the natural attitude (or in common sense), the life world is still far from being fully coherent (as will presently become apparent); so that perceptual consciousness itself becomes an object to a further stage of conscious reflection, to be considered below.

Two questions now arise: How do we know all this to be true? What is the criterion of veridicality? The misgiving that the view so far outlined is too subjective has already been countered in a preliminary fashion, and has yet to be further addressed. Admittedly, we cannot get outside our own consciousness, but consciousness is itself the activity of ordering the contents of sentience; and that is, as we have asserted, a unified whole of feeling which, in the very course of the process of organization, is revealed as the focused registration of a world external to the sentient body. Apart from this organization nothing whatever becomes intelligible, so its product cannot be impugned as false *in toto* for it alone can provide the criterion of truth.

The belief in the reality of the life-world is immediate and innate, its initial justification is primarily pragmatic—the consequent success, when reliance rests upon it, of the organism's self-maintenance. But such practical satisfaction is (as Plato saw long ago) not sufficient in itself as a criterion of truth. It in its turn depends on the standard of value adopted, which again draws upon judgements of fact concerning one's own (human) nature and one's relation to the world at large. These can only be judged by some other criterion of truth than pragmatic success, for that depends upon them. Moreover, if the organism and its self-maintenance were themselves no more than items in a perceived situation, judgements of success or failure would be as subjective as any other. The pragmatic criterion is largely a matter of coherence between the perceived situation, the experience of wanting, and the achieved result of action. The ultimate justification of the belief in the reality of the external world demands more than this. It requires reflection and criticism, of which there is further discussion to come.

The criterion of truth is always the same. It is the coherence of the experience as a whole, and this is implicit even in pragmatism. Similarly, the contention that to be true, theory must agree with the facts, if it is to mean anything credible can only express the requirement that interpretation of what is perceived or observed must not give rise to contradiction. If theory does not agree with practice, then obviously either the first must be wrong or the second ill-advised, or both. Whether it agrees with the facts will depend in part how "the facts" are interpreted, and that itself involves theory, so that ultimate coherence is all that we can appeal to.

Detection of illusion and hallucination relies on the same criterion. Tests of the illusory object by further related perception give rise to incoherencies that reveal its falsity. The mirage, as approached, disappears; what seems to make contact with the apparent water causes no splash (an inconsistency with past experience), and so forth; the Muller-Lyer illusion is exposed as false by measurement—another comparative perception. In applying such tests, we remain within the perceptual horizon. There can be no appeal to *unperceived* "fact"; consequently, the only criterion can be the mutual harmony and concord within that experience, not between it and something else—something external or independent.

From the very first, at the level of simple perception, what is perceived is what the mind finds coherent, both within itself (as *Gestalt*) and with its context. And when the percept is illusory, in whatever way, it is recognized as such only when contradiction is found between what appears and the context in which the mind is constrained to include it. The psychologists to whom we have referred all give testimony to this fact. Their investigations, in particular of constancy phenomena, of the perception of size in relation to distance, and of the perception of incongruity, give copious evidence of the function of coherence as the subconsciously accepted criterion not simply of veridicality, but of the actual nature of the object perceived.[24]

SCIENTIFIC OBSERVATION

The difference between common perception and scientific observation is not one of kind but only of degree of sophistication. Before the days of telescopes and microscopes, the astronomer and the naturalist took note simply of what they normally perceived, and differed from other people only in what especially interested them and the way they thought about it. They noticed something odd, something incoherent, about the appearances, which prompted them to ask questions, to frame tentative answers, and then to observe more carefully and attentively, seeking evidence to support those answers. The search for evidence in support of hypotheses leads scientists to devise experiments. These are no more than ways of observing under conditions specially arranged to test whether the anticipated results will be observed. The ordinary person does not perform experiments unless perception is impeded or something puzzling is presented to the viewer. Only when questions arise are we prompted to experiment. And questions arise when we are presented with appearances that are in mutual conflict.

If contemporary psychological theories are to be respected, there is then no difference in principle between ordinary perception and scientific observation. Both are active efforts to discern presented objects by a subject framing hypotheses and trying to confirm them by correlating evidence for and against. In the first case the process is largely subconscious and what H. H. Price called pre-judicial (although it is nevertheless incipient judgement). In the second, it is deliberate and explicit.

Sensory stimuli are invariably ambiguous. An ornament in a wallpaper pattern, a spotlight playing obliquely on a level surface, the shiny lid of a jam jar seen at an angle, or a shaft of sunlight piercing clouds and illuminating a distant field may all equally appear to a perceiver as a white elliptical patch. Which of these it represents only the context coupled with past experience will reveal. And the process through which these two mediating influences work is interpretive thinking. But again, at the conscious level conflicts and incoherences occur that raise new problems.

In ancient times the movement of the planets, contradicting that of other heavenly bodies, raised questions that excited thinkers to frame hypotheses and excogitate answers. Much earlier, problems of navigation or of agricultural practice induced people more carefully to observe the heavens and the movement of the sun and the stars. Thus was the science of astronomy established, and thus is scientific observation always motivated—by practical needs and conflicting appearances. Inquiry is prompted by contradictions first within common experience. These raise questions to which thinkers propose tentative answers, and then proceed to investigate further by more careful observation. But the result of the observation is not immediate nor always (if ever) conclusive by itself. It has to be interpreted and fitted into a system of thought that is coherent and comprehensive.

Common perception involves the fitting of sensuous cues into structural schemata that I have assimilated to concepts and that together build up into a commonly accepted, if only roughly coherent, picture of the experienced world. Science proceeds, likewise, under the aegis of conceptual schemes which together make up a more systematic world-picture. Thomas Kuhn calls these schemes, somewhat infelicitously, "paradigms." And science progresses by applying these concepts to puzzling phenomena that demand explanation. When, as it proceeds, new contradictions emerge, the accepted theories are modified to preserve coherence. To speak of theories agreeing or failing to agree with fact is inept, because the observed fact (as is now generally recognized) is itself an interpretation largely dictated by theory (it is theory-laden). What is required is that theory and observation taken together form a coherent system within the conceptual scheme currently accepted.

Kuhn's account of this procedure is, in the main, sound. It proceeds, he avers, through a period of "puzzle-solving" or "normal science" within the orbit of the prevailing "paradigm"; but when inconsistencies and contradictions, hitherto set aside or overlooked in the expectation that future progress will eliminate them, accumulate to the point at which research is drastically frustrated, new hypotheses proliferate until a new paradigm (a new conceptual scheme) emerges which resolves the problems that were intractable under the one hitherto accepted. Kuhn errs, however, in his insistence on the incommensurability of different paradigms, due to his failure to appreciate the dialectical logic of revolutionary change and the consequent continuity of scientific advance. I have discussed these issues elsewhere;[25] here other consequences of the theory are more relevant.

The contemporary recognition that scientific observation is theory-laden came with N. R. Hanson's book *Patterns of Discovery*, followed shortly afterwards by Kuhn's declaration that what scientists observe is what the "paradigm" dictates. Attention is always selective, and what guides it is interest, on the one hand, and previous knowledge, on the other. What is perceived, therefore, is partly what is expected and partly what is sought, never simply what is there. A vast amount is merely overlooked because we are not interested in it and do not attend to it, yet more is simply not perceived because it is not credited as possible. In consequence, the untrained eye simply misses what the expert takes in at a glance, the untrained ear fails to distinguish what is palpable to the practised musician. Similarly, the physicist will see the trail of a positron in a Wilson cloud chamber, where a layman will see only rapidly changing white streaks against a dark background. Or a surgeon will detect an anatomical abnormality recorded on an X-ray film, which to the patient reveals only shadowy unrecognizable shapes.

Yet even the scientist will miss what the current conceptual scheme does not accommodate. Darwin, in his autobiography, recalls how, on a geological expedition, neither he nor Sedgewick observed the striation on the rocks or the balanced boulders, so patently indicative of glacial effects. Twenty times before Herschel discovered Uranus, the planet had been sighted by other astronomers who had failed, even when its movement was apparent, to see it as a planet.

The conclusions to which all this leads are, first, that scientific observation is continuous with and in principle no different from ordinary perception, in that both are intrinsically interpretive. Both are shaped and informed by *a priori* concepts. Secondly, scientific observation is continuous with common sense, as raising to a higher degree of systematization what in common sense is already the experience of an ordered world.

The perceptual world is, like the mass of primitive sentience from which it emerges, a whole, but now more articulately differentiated and more highly organized; and science continues the process of organization to produce a world-picture at a still higher level of systematic thinking. The scale of forms extends continuously from physical wholes to chemical, from chemical unities to organism, from organism to psychical field, from feeling to perceptual experience, and from the common-sense world to the scientific world-view.

It is the same totality throughout, in different phases of self-articulation. The higher in the scale the form (or provisional whole) is, the more adequate it is to the universal principle of organization immanent in every phase. The scientific world-view, therefore, will be what Hegel called "the truth of" nature, at least as a succeeding dialectical phase. What is observed in science is thus the actualization at the conscious level of what was merely implicit at the physico-chemical or the organic level. While they are each, in their appropriate degree, real, the more truly real, the more concretely actualized (and not, as so often held, the more abstract, even though abstraction is an essential step in scientific methodology) is the rationally and systematically conceived world-picture offered by science. In conclusion, let us review the perceptual scale of forms. Sentience, as we saw, is a self-contained whole of feeling. It differentiates into object and background, the structure of attention-levels forming a specific whole. Figure-and-ground emerges as a primary *Gestalt*; then the material object in its spatio-temporal schema; schemata (wholes in themselves) order such objects into situations, each giving a vista, which is a further whole—in art, a pictured scene or a scenario—and this resides within the life-world, the universal perceptual horizon. A refinement of all these and continuous with them is the scientific conceptual scheme, differentiating into hypotheses and theories and culminating in the scientific world-view. But these belong more properly to the new and yet more critical level of self-reflection which is the subject of the next chapter.

Notes

1. This is a misconception into which some physiologists and psychologists, as well as many philosophers besides John Locke, have fallen. Cf. Kenneth Craik, *The Nature of Explanation* (Cambridge University Press, Cambridge, 1952), and E. L. Hutton, "The Relation of Mind and Matter to Personality," in *Perspectives in Neuro-Psychiatry*, ed. D. Richter (H. K. Lewis, London, 1950), among others.
2. John Locke, *Essay concerning Human Understanding*, II, viii, 12.
3. Causal and representative theories of perception have been extensively discussed and refuted in the literature. I have myself discussed them in more detail, from Locke to Russell, in *Nature, Mind and Modern Science* (George Allen and Unwin, London, 1954, 1968), and with reference to physiological and cybernetic theories in *The Foundations of Metaphysics in Science* (George Allen and Unwin, London, 1965; reprinted by the University Press of America, 1983), Ch. XIX.
4. H. H. Price, "Appearing and Appearances," *American Philosophical Quarterly*, 1, no. 1 (1964).
5. Cf. Immanuel Kant, *Kritik der reinen Vernunft*, A99. Such cogent criticisms are to be found in B. Blanshard, *The Nature of Thought* (George Allen and Unwin, London, 1939, 1948), and H. H. Joachim, *Logical Studies* (Oxford University Press, Oxford, 1948); for my own discussion see *Nature, Mind and Modern Science*, Pt. IVB, and *Hypothesis and Perception*, Ch. VIII.
6. Cf. G. A. Paul, "Is There a Problem about Sense-data," *Proceedings of the Aristotelian Society*, Supplementary Volume, XV (1936).
7. "The Myth of Sense-Data," *Proceedings of the Aristotelian Society*, XLV (1944–45).
8. "The Theory of Appearing," in *Philosophical Analysis*, ed. Max Black (Prentice Hall, Englewood Cliffs, NJ, 1963).
9. Cf. Adelbert Ames, Jr., "Visual Perception and the Rotating Trapezoid Window," *Psychological Monographs*, 63, no. 7 (1951); "Reconsideration of the Origin and Nature of Perception," in *Vision and Action*, ed. S. Katner (Rutgers University Press, New Brunswick, NJ, 1953); and Errol E. Harris, *The Foundations of Metaphysics in Science*, pp. 405–406.
10. John Austin, *Sense and Sensibilia* (Clarendon Press, 1964), p. 8.
11. Cf. E. H. Gombrich, *Art and Illusion* (Phaidon Press, London, 1959–62; Bollingen Series, New York, 1951).
12. Russell Brain, *Mind, Perception and Science* (Blackwell, Oxford, and C. C. Thomas, Springfield, IL, 1951), p. 32. Cf. also H. Werner and S. Wapner, "Sensory-tonic Field Theory of Perception," in *Perception and Personality*, ed. J. S. Bruner and D. Krech (Duke University Press, Durham, NC, 1949); and G. L. Freeman, "The Problem of Set," *American Journal of Psychology*, 52 (1939), and *The Energetics of Human Behavior* (Cornell University Press, Ithaca, NY, 1948).
13. Cf. my discussion in *Nature, Mind and Modern Science*, Ch. IX.

14. R. Borger and A. E. M. Seaborne, *The Psychology of Learning* (Penguin Books, Harmondsworth, 1966), p. 118.
15. E. H. Gombrich, *Art and Illusion*, p. 45.
16. Cf. *inter alia*, B. Blanshard, *The Nature of Thought* (George Allen and Unwin, London, 1948); C. A. Campbell, "The Mind's Involvement with Objects," in J. Scher, ed., *Theories of the Mind* (Macmillan, London and New York, 1962). See also my discussion in *Hypothesis and Perception*, Ch. VIII, and "Blanshard on Perception and Free Ideas," *The Philosophy of Brand Blanshard*, ed. Paul Schilpp (Open Court, La Salle, IL, 1980). For a critique of this theory see R. J. Hirst, *Problems of Perception* (George Allen and Unwin, London, 1959).
17. Cf. G. Ryle, "Sensation," in *Contemporary British Philosophy*, Third Series (George Allen and Unwin, London, 1956); A. M. Quinton, "The Problem of Perception," *Mind*, LXIV (1955), pp. 28–51, and "Perception and Thinking," *Proceedings of the Aristotelian Society*, Supplement, XLII (1968); D. M. Armstrong, *Perception and the Physical World* (Routledge and Kegan Paul, London; Humanities Press, Atlantic Highlands, 1961); M. Merleau-Ponty, *The Primacy of Perception* (Northwestern University Press, Evanston, 1964).
18. Cf. my *Hypothesis and Perception*, Ch. VIII, and *Foundations of Metaphysics in Science*, Ch. XX; F. C. Bartlett, *Remembering* (Cambridge University Press, Cambridge, 1961), Ch. II; M. D. Vernon, *The Psychology of Perception* (Penguin Books, Harmondsworth, 1963–68), and *Experiments in Visual Perception*, ed. Vernon (Penguin Books, Harmondsworth, 1966, 1970); R. R. Blake and G. V. Ramsey, eds., *Perception—an Approach to Personality* (Ronald Press Co., New York, 1957), Ch. 5; J. H. Rohrer and M. Sherif, eds., *Social Psychology at the Crossroads* (Ayer, New York, 1951), Ch. 10.
19. Cf. Blake and Ramsey, *Perception*.
20. Cf. F. C. Bartlett, *Remembering*, Ch. II.
21. Cf. E. Husserl, *Experience and Judgement*, translated by J. S. Churchill and K. Ameriks (Northwestern University Press, Evanston, 1973), § 12, pp. 54f.
22. Cf. references given in notes 16, 17, and 18, above.
23. Cf. Ch. 2, above.
24. Cf. W. H. Ittleson, "Size as a Cue to Distance: Static Localization," *American Journal of Psychology*, 64 (1951); J. S. Bruner and L. Postman, "On the Perception of Incongruity," *Journal of Personality*, 18 (1949); R. H. Thouless, "Phenomenal Regression to the Real Object," *British Journal of Psychology*, 21 (1931); C. R. Borresen and W. H. Lichte, "Shape-Constancy: Dependence upon Stimulus Familiarity," *Journal of Experimental Psychology*, 63 (1962); E. E. Harris, "Coherence and Its Critics," *Idealistic Studies*, V, no. 3 (1975).
25. Cf. Thomas Kuhn, *The Structure of Scientific Revolutions* (University of Chicago Press, Chicago and London, 1962); and my *Hypothesis and Perception*, Ch. VII, "Dialectic and Scientific Method," *Idealistic Studies*, III, 1 (1973), republished as "The Dialectical Structure of Scientific Thinking," in *Hegel and the Sciences*, ed. R. S. Cohen and M. W. Wartofsky (Reidel, Dordrecht and Boston, 1984).

10
Reflection

CRITICAL TRANSITIONS

In the scale of forms that constitutes the self-differentiation of the cosmic order there are two critical transitions. The first is from the physico-chemical to the metabolic, marking the emergence of life. The second is from the sentient and perceptive to the fully self-conscious and reflective. Neither of these is abrupt or unheralded. Life is foreshadowed by crystalline and organic molecular structures; reflection is preceded by immediate perception. But the crucial awakening is that of reflective deliberation, because here for the first time the universal principle of organization, as such, begins to become explicitly aware of itself as reason.

The universal principle is dynamically self-specifying. It manifests itself first in a physical universe, then in an organic totality, and subsequently in a known world (or noösphere). Only then is its concrete potential fully actualized, because only then does its systematic structure become explicit. At this stage, and in the process of its occurrence, it becomes aware of itself as conscious subject, reflective upon itself, upon its own experience of itself and of the world. Its self-transcendent awareness becomes active in criticism and evaluation in morality and intellect, the level on which it is explicitly capable of transcending its own finite limits to become apprised of, and to comprehend, its own infinite scope and potential.

Physical particles in mutual relation, so far as they are individual entities, are each relatively self-contained, and their subjection to external influences is imposed upon them from without. They operate blindly, and even although the organizing principle of the physical system is immanent in every element, determining interaction between them, the structure of the system is never *explicitly* expressed in physical work or movement, but is only evident to us as theorists.

The generation of life marks a new departure and a further advance in centreity. Metabolism, in intertwined cycles of catalytic chemistry, achieves the self-maintenance of the organic system. But not yet is the interrelation of parts explicit, nor the organic holism *für sich*. It is still only immanent; it may be felt in some obscure fashion (for all we can tell), but we can be

139

reasonably sure even of this only after the living creature has evolved to some considerable extent. Not until the unifying principle becomes aware of itself do the relations between the parts of the whole become explicit—in short, not before the emergence of consciousness and its transcendental subject.

SELF-TRANSCENDENCE OF CONSCIOUSNESS

The miracle of consciousness is self-transcendence. It is primarily the apprehension of relations, and no relation can be grasped within the limits of any one of the terms. The conscious subject must be able to stand aloof from all the terms and to hold the entire complex before its mental view in order to become aware of it as a whole. But when, as always, the subject is itself one of the terms, this requires not merely self-consciousness but also self-transcendence, which self-consciousness necessarily involves; for no self can be aware of its own identity apart from an intentional object, and (as Fichte and his immediate followers well knew) to become an object to itself the subject must divide and limit itself: it must, as it were, project itself beyond itself and alienate itself from itself. In short, it must transcend itself.

All consciousness is at least inchoate self-consciousness, because to be aware of an object is (as Spinoza insisted) to be aware that one is aware of the object. So consciousness of any object is implicitly and immediately consciousness of oneself. Hence Descartes's confident declaration: *Cogito ergo sum* (where *cogito* means "I am conscious").

Moreover, to be conscious of an object is to cognize it in a context both spatial and temporal. But to be aware of a temporal context is at once to remember and to anticipate. What is remembered or anticipated is not occurring now, yet it is remembered or anticipated now. Whatever is present in consciousness, and so related to the event remembered or anticipated that the subject is now conscious of the past or future occurrence, is at once distinct from that event yet identified with it, which cannot be unless the consciousness transcends the present moment and the present representation. For no image that I now entertain can *be* the remembered (or anticipated) object, nor can I recognize it as an image of that, or of anything, unless I am at the same time somehow aware of the object of which it is an image. But, in the instance considered, that object is not present, and so can be apprehended only by a transcendent consciousness.

All consciousness of time involves such transcendence, because the succession of events can be apprehended as a succession only if the series is grasped as a whole, which means that the apprehending subject can never be confined to any one event, past, present, or future. It must be transcendent above, or beyond, all of them. Similarly, although spatial expanse appears to sight as present in a single instant (at least, within limits), which it is not to the other senses (apart from inference and imaginative construction), the interlocking of perspective views involved in the perception of solids and their precise location in space requires their interrelation and, so, the transcendence of all such perspectives by the thinking subject who becomes aware of those relations. Much the same is true of the perception of space through touch, which requires the integration of numerous touch sensations, of pressure (what H. H. Price called "premencies"), heat, cold, and the like, sometimes concurrent but often successive; and the perceiving subject can be wholly identified with none of them but must grasp them in mutual relation.

The concept of a transcendental subject has given philosophers (and some psychologists) trouble ever since it was proposed by Kant as indispensable to any experience whatsoever. In the Transcendental Deduction, Kant demands it—the agent of transcendental *a priori* synthesis of the manifold of sense, the transcendental unity of apperception—as the condition of the possibility of any object of experience whatever. But then, in the Transcendental Dialectic, he shows that the *ego* cannot itself be known as an object of experience, for instance, as a substance (or soul), because that would require its being brought under the categories of the understanding, of which it is itself the *a priori* source. Consequently, the attempts of reason (of earlier metaphysics) to conceive the soul as a substance involves paralogism. What one perceives as oneself is what Kant calls the empirical self, which is little more than the conscious stream of inner sense. But here there is no identifiable *ego*, for that never appears, and cannot appear, in inner perception.

Hume had reached a similar conclusion, discovering no impression in his experience that corresponded to the self. That, he maintained, was merely a bundle of impressions and ideas, appearing successively in consciousness with incredible rapidity, without any "self" ever being presented as such, as an identifiable object.

Phenomenologists, following Husserl, have again found need to postulate a transcendental *ego* and then again found it difficult (as Husserl was never successfully able) to make its relation to the empirical self properly intelligible. Heidegger and Merleau-Ponty wrestle with the problem, but, as I have tried elsewhere to show, they fail to dispose of it satisfactorily, or without at least implicit self-contradiction; while Michel Foucault, in repudiating the transcendental subject altogether, becomes hopelessly involved in self-refuting methodology. The arguments that support these criticisms I have set out in *The Reality of Time* and I shall not repeat them here.[1] The contemporary psychologist Thomas Natsoulas, in discussions of conscious perception, finds similar difficulties with the concept of an "I." Natsoulas supports the positions expounded by Wittgenstein and by Gilbert Ryle, who seek to demonstrate the illegitimacy of the concept, and he argues against the views of G. G. Globus, who strongly defends the belief in, and claims the direct experience of, the "I." Wittgenstein and Ryle, by and large, adopt an attitude that is generally behaviouristic, against which some strictures have already been drawn above; while Globus argues from introspection.[2]

Natsoulas complains of the absence of empirical grounds for an "I" and concludes, in much the same way as does Hume, that the idea is illusory. "I do not experience 'I' as I do my body," he writes, "or even as an imaginary dragon . . ."[3] Kant would have agreed that I do not experience "I" as an empirical object, so that there can, of course, be no "empirical grounds" of that sort for its existence. Yet Kant had established beyond dispute that without a transcendental synthetic unity of apperception, no experience of any objects, no empirical evidence of anything at all, is possible.

"[Our] awarenesses are happenings in our nervous system," Natsoulas asserts, "and can occur without our being (erroneously) aware that an 'I' does them . . ."[4] Nevertheless, Natsoulas, as he presents his arguments, continues to refer to himself as "I," and we may wonder to what he is referring when he does so. Could it be to an illusion?

Against the neural identity theory of consciousness, I have argued, I hope conclusively, that our awarenesses are not just "happenings in our nervous system." What they are certainly does involve our bodies, but as organic wholes, and the awareness of the "happenings" is the form taken by that wholeness at a high degree of integration. It is the form of feeling, which becomes consciousness when it is organized by attention and judgement identifying, distinguishing, and relating objects. But no such judgements and no awareness of relations could occur without a subject who transcends the single terms of the relationships. The *ego* is not any one of these terms. It is *the whole* come to consciousness of itself as "I." As such, it can and does distinguish itself from its objects, including its body, in which the neural happenings occur, in order to be aware of them as physiological. Indeed, I am not a separate or separable entity from my body. I am identical with it, or rather I am its identity as a functioning whole—the self-cognizant form of the principle of unity and organization immanent in it. The body can, even when injured within certain limits, function as an integral whole, but if it fails to do so, I cease to be conscious of myself or of anything at all. As the self-awareness of the universal principle of wholeness in the body, the "I" has become transcendent over the objects of its consciousness. To use Hegel's term, it overreaches them (*übergreift*). As objects they are its other, yet it remains identical with them, the content of its sentient experience.

SELF-REFLECTION

Thus consciousness necessarily implies a transcendental subject, but it does not become aware of itself until a late stage of development. The evidence of conscious perception informing the behaviour of birds and mammals is virtually overwhelming.[5] But only among the great apes is there any sign of self-consciousness proper. Chimpanzees have been taught to indicate themselves in sign language, but beyond that, and prior to human personality, there is scant evidence of the ability to refer to the self as "I."

Awareness of self is reflective consciousness, in which the subject becomes its own object. To be aware of oneself and one's condition, to be able to grasp the relation between oneself and others (both persons and things), to perceive the situation in which one is involved and how one is liable to be affected by it is to be capable of reflection and deliberation. Nothing less than this is required for critical assessment of the situation and evaluation of its prospects. (This, no doubt, was what Heidegger had in mind when he wrote of choosing among one's envisaged *Seinkönnen*.) Reflection, as it is implied in self-consciousness, is thus necessary to criticism and evaluation, and so to deliberate thought and action. Reflection, which leads to deliberation and criticism, is therefore essential to all morals and politics. It is what gives rise to questioning and wonder, and so is equally essential to all science and philosophy, and because it is the root of the awareness of the distinction between the finite and the infinite, also to religion.

Only at the stage of reflection can there be any science. Only on this level do human beings notice anomalies in their life-world and raise questions, construct hypotheses, and embark upon investigations to confirm or disprove them. If, then, what scientists discover is affected by the intelligent and reflective character of their thinking, so that the content of what they find is the consequence of their capacity to investigate, if, in short, the Anthropic Principle holds, it can only be because scientists have reached the level of self-reflection, and only at that level can the principle be recognized. But this self-reflection is the outcome of the bringing to self-consciousness of the organizing principle of the whole through the process of its own self-specification. Consequently, its self-awareness is the awareness of that process, is its knowledge of its own concrescent nature and the way in which it must specify its universality; in other words, it is the knowledge of the world of nature. What the scientist discovers, therefore, is the entire natural process that brings him and his scientific thinking into being, and which does so for no other reason than that its coming into being is required by the systematic unity and wholeness of the universe itself. It is precisely this that is signified by the Anthropic Principle.

NEWTONIAN SCIENCE AND THE UNOBSERVED OBSERVER

Consideration of morality, politics, and religion must be deferred for the present. Our immediate concern is with science. It has already become apparent that scientific inquiry arises from the occurrence in perceptual consciousness of contradictions that create puzzlement and raise questions, the attempt to answer which leads to speculation and then to systematic observation. Although obstruction of action and some consequent feeling of frustration may occur before the reflective level of consciousness has been reached, puzzlement and questioning cannot. These are the fruit of reflection because they involve critical assessment and deliberation. Science begins in wonder and the interrogation of nature, which presuppose reflection, so without that there can be no science, and without science knowledge of the world remains in its infancy.

Science proceeds within the confines of a conceptual scheme as long as conflicts do not arise within it in the course of its specification. When they do, and when they impede the progress of investigation, modifications are called for to restore harmony in the "constellation" of presuppositions (as Collingwood called them)—modifications that may well give rise to a scientific revolution.

Such revolutions have occurred at various stages in the history of science, perhaps the most often discussed being the Copernican, which in its Newtonian development produced the mechanistic world-view. But the Newtonian instauration provided no mechanics of the mind, and the celestial mechanics made no room for consciousness. The scientific observer viewed the world, as it were, from the outside, and within that world no provision was made for any consciousness. The observer, therefore, remained outside the purview of investigation and was presumed to make no difference to its results. In philosophy (the further fruit of reflection) this led to materialism, as we see it in Hobbes and Gassendi, and to dualism, as we find it in Descartes and Locke. Materialism is refuted by the considerations that neither that nor any other theory would be possible without conscious self-reflection, and that no material process is self-reflective. Dualism, whether rationalist or empiricist, collapses for the like reason: its failure to provide a viable theory of knowledge, of how the mind, in its separation from matter, can encompass a representation of an external world, how the world can get into the mind, or how a mechanical material process can be miraculously converted into a cognitive awareness. Meanwhile, the scientific disregard of the observer leaves science itself beyond the reach of scientific explanation. Any attempt to understand what science is doing is hamstrung, and any account of it is falsified, if its self-reflective character is overlooked.

THE OBSERVER REINSTATED

The impasse which, at the end of the 19th century, obstructed scientific progress within the Newtonian "paradigm" could be broken by nothing less than a new revolution, which required a more holistic approach (although holism was dimly adumbrated even in classical physics); and this came with relativity and quantum theories. Neither of these could disregard the observer. For relativity, the relative velocity of the observer determines the value of every measurement, and for quantum theory, the observer and the measurement of specific quantities have become inseparable from the very actuality of elementary particles. It is now being suggested that reflective awareness, in the guise of observation and interpretation, is constitutive of the very being of the universe.[6]

SUBJECTIVISM

Here we encounter the menace of subjectivism, which takes two main forms: one Berkeleyan, threatening to dispense altogether with independent physical reality, the other Kantian, reducing all knowledge of the physical world to a phenomenalism that leaves the reality beyond our ken as an unknowable thing-in-itself.

(1) The Copenhagen interpretation of the quantum theory (due to Niels Bohr) links the observer and his choice of experiment inseparably with the observed phenomenon. Two considerations support this position: first, the Schrödinger wave function, according to his own interpretation, measures the charge density of the electron in a system described by the Schrödinger equation. The wave function describing a beam of electrons impinging upon a photographic plate is therefore widely extended in space, although each electron falls at a precisely defined point. Schrödinger interpreted this impact as the instantaneous collapse of the charge-spread to a definite point. But that would imply the instantaneous transfer of information over an extended region of space (i.e., faster than light) forbidden by the special theory of relativity. Secondly, the Heisenberg indeterminacy principle makes it impossible to measure conjugate properties of a particle simultaneously. Bohr therefore maintained that the indeterminate properties do not exist as physical realities but come into existence only when the measurement is performed that reveals their precise values to the observer.

Einstein and others opposed this interpretation, maintaining that particles with their properties were real independently of the observation, and that the indeterminacy might be accounted for by presuming the existence of hidden variables, but Bell's Theorem establishes the impossibility of explaining instantaneous transfer of information on the assumption of such hidden variables. Further, it was not feasible to assume that the collapse of the wave function occurred simply within the measuring instrument, for the wave function could be written so as to cover the state of the apparatus as a whole, as the Schrödinger cat paradox illustrates. One must, therefore, so it seems, conclude that the physical reality comes into being only with its actual apprehension by the observer.

Eugene Wigner has argued with great cogency that the collapse of the wave function can occur only with the observer's consciousness of the result of the measurement; and this, if generalized, makes the reality of the physical entity dependent upon its being perceived: *esse est percipi*. Thus, we seem committed to a Berkeleyan idealism, which we know cannot ultimately withstand criticism, in that it would lead inevitably to solipsism and consequent self-contradiction because ultimately only the conscious subject (the observer) has warrant of existence—but without a not-self, no self can become aware of its own identity.

At the same time, Henry Stapp's demonstrations have shown that the result of the EPR experiment, taken together with Bell's Theorem (the source of the original *aporia*), makes inevitable the assumption of faster-than-light influences that are not signals, giving additional strong evidence of the indivisible unity of the physical universe. We are thus brought to realize that the unity of the universe and the apparent dependence of physical reality upon subjective experience are two aspects of a single fact.

If the universe is an indivisible whole, and as such must by its very nature be complete, and if, as has been argued, the completion of a whole necessarily involves its being brought to consciousness, the danger of falling into solipsism is averted. For, although hidden variables have been ruled out, the indeterminate properties of particles are admitted by the Copenhagen theorists to be latent, or potential, before they are observed. Now, in the self-specification of the universal principle of organization that governs a whole, the prior phases in the scale of forms are less adequate to its generic nature than the subsequent phases; and it has already been disclosed that the organizing principle is merely implicit (not fully manifested) at the physical and biological levels, coming overtly to light only with the emergence of consciousness. It is therefore wholly concordant with the theory here being expounded that the actualization of what is potential at the physical and biological levels should await the activity of observation and the efflorescence of knowledge. This in no way precludes the prior reality of the physical and the biological, because the very experience of a physical and biological world as an indivisible systematic whole implies and necessitates the self-differentiation of that whole as a scale of forms, in the more elementary of which what emerges at later stages is already implicit.

Objection may be raised on the ground that the crucial issue has not really been addressed. In an earlier chapter, the Participatory Anthropic Principle (which is what we are really concerned with here) was assailed for committing a *hysteron proteron*. The reality of elementary particles is restricted to the act of observation by means of instruments that are themselves composed of multitudes of such particles, and by observers who have evolved from organic species similarly composed. Thus the reality of the elements is made to depend on the activity of that to which they are elementary. Raising the entire physical cosmos to the conscious level through the mediation of an evolutionary scale does not seem to affect this paradoxical anthropic claim, nor to address the ensuing problem.

The objection might be met by pointing out that the quantum theory (like any other scientific theory) could be conceived only at the level of reflective thought, the prior development to which must, therefore, unavoidably be presumed. The existence of both macroscopic and microscopic physical entities is thus established, and all that is at issue is the interpretation of the mathematical formulation of quantum mechanics. If the Copenhagen interpretation is preferred, the quantum system must be regarded as latent, or merely potential, until determined by measurement; and that is compatible with the philosophical position here being advocated. If, however, this submission fails to satisfy, the alternative Bohm-Hiley interpretation might be adopted, giving ontological status to particles moving under the control of a quantum field. This also conforms to the philosophical concept because it maintains the holistic conception of a physical reality that requires dialectical explication, such as has been set out. Then the system only becomes clearly explicit when cognized and expressed in scientific terms, so that the Participatory Anthropic Principle still holds.

(2) Quantum theory, in a somewhat different interpretation, has inspired another form of subjectivism. The principle of indeterminacy and the probabilistic character of the Psi-function have led some reflective scientists to conclude that physical reality may well be a chaotic welter of energy on which the appearance of order supervenes only because we impose upon it a stochastical mathematic. Quantum mechanics, it is held, has abolished causality, and for the quantum physicist, it might be said, there can be no causal laws. What we take to be laws of nature are simply statistical estimates of what will probably happen in the general run. Accordingly, say those who follow this line of thought, our conception of the universe is a subjective fabrication, constructed to serve our practical needs and to satisfy our penchant for orderly conception. The actual nature of the reality, if we cannot presume that it is just sheer chaos, we can never discover.

In several earlier writings I have tried to show that this line of argument conceals a fatal inconsistency, so long as it is presumed that our minds are natural products.[7] For, if they are, there must be some principle of order in nature which can generate agencies capable of imposing order on experience. How else could such agencies emerge from an allegedly chaotic matrix? The only escape from this implication would entail abandonment of the belief in the continuity of evolution and renunciation of all connection between the human mind and physical nature. We should have to assume that mentality could occur only by virtue of some kind of special creation. Then, at best, we should return to an unmanageable dualism, like that which dogged the heels of Renaissance science and its attendant philosophies. Indeed, it may well be that the type of approach under scrutiny is a relic of that outmoded paradigm. At worst, this position would be committed to an extreme subjectivism that has already proved to be unsustainable. For if the reality is in principle unknowable, we are left solely with what our own consciousness presents.

Professor Edward Harrison has adopted a similar view, apparently from historicist considerations. In the history of science, he points out, a number of world-views have arisen in succession, each to be rejected by subsequent thinkers, often as palpably false and ridiculous. No view of the universe that we can frame, he contends, has any sustainable claim to truth. Each successive world-picture is no more than a mask that hides the true features of the reality, but reality itself remains for ever unknowable.[8]

Such historical relativism brands science as little better than fantasy, by dissociating it entirely from the reality, which is held to be beyond its reach, and of which, in such circumstances, it could give no credible account that would ever rank as knowledge. Thus the very name "science" would be belied. But if, on the contrary, the history of science is recognized as a dialectical succession of provisional conceptual schemes, unfolding as a scale of forms, in which each progressively more adequately explicates a conceptual whole;[9] and if each conceptual scheme is itself a particular stage in the self-discovery of the actual world, we can accept everything that Professor Harrison sees in that history without being abandoned to an implied and inexorable scepticism.

The succession of scientific revolutions deploys a series of world-views that increase from each to the next in coherence and unity. The cosmic whole, differentiating itself and bringing itself to fruition in self-awareness, is depicted in these *Weltanschauungen*, but never completely or in adequately detailed systematicity. Our scientific theories are never more than approximations. The vast complexity of the real exceeds the scope of the algorithms mathematicians have, as yet, devised, so that, in order to cope with it, physicists often have to ignore the finer details, to generalize, and to approximate. Even so, they continually find ways to increase the accuracy of their measurements and predictions and with them the comprehensiveness and coherence of their theories. But no world-view is ever completely satisfactory. New puzzles constantly arise, and unresolved problems remain. It is these that keep the scientist in business and persistently stimulate continuing research. But the comprehensive coherency of the world-view guarantees, according to its degree, the reality of its object, without committing the conceiver to any Kantian variety of scepticism with respect to the reality of the actual world.

THE NOÖSPHERE

The universal principle of organization immanent in all things manifests itself in a cosmic pattern, in which it is particularized in successive wholes, constituting the scale that has so far been traced. The physical universe, by self-enfoldment of its elementary parts, becomes chemical; and then, through further self-enfoldment, it is sublated in an organic world. That again, by similar self-enfoldment and complexification, internalizes the whole in sentience, and then articulates it as an experienced world, which is further systematized by scientific analysis and synthesis. The physical and biological worlds are thus transfigured into the noösphere; and at that level the whole can be viewed as one unbroken series of mutually continuous wholes. The noösphere can now be seen as actuallizing more fully the implicit potentialities of the biosphere and its physico-chemical prerequisites. Within this purview, the paradoxes of the quantum theory, of Schrödinger's cat and the threatened subjectivism of Wigner's interpretation are resolvable, and the PAP can be consistently maintained.

The noösphere, however, is by no means confined to the mathematical and empirical sciences. While it is open and available to any mind, the scientific *Weltbild* is neither produced by, nor is it the exclusive possession of, any one individual. To develop, it requires a scientific community who generate it through cooperation, teamwork, and discussion. But this community is itself parasitic on (as well as contributory to) the wider organized social whole, of which Aristotle wrote that it "comes into existence, originating in the bare needs of life, and continuing in existence for the sake of a good life."[10]

In the matrix of this society the noösphere is engendered at many levels and its development ramifies in many directions. Out of this soil grows morality. In the social *milieu* individual persons are so interdependent that their lives and welfare depend on the common observance of rules. At the human level, social and individual needs generate crafts and arts, the specialization of functions, the division of labour, and the exchange of products. Social structure becomes more elaborate, demanding more formal regulation, and a political system develops. None of this can be divorced from a social awareness of tradition and so of history, while, under the aegis of the social order, science and technology emerge and the reflective urge of consciousness fulfils itself in philosophy and religion.

Although these facets of the noösphere are related in dialectical order, their mutual dialectical involvement is far more complex than that of physical or biological natures. Because they are the fruits of reflective self-consciousness, they partake of its transcendent character, each phase being at once the whole and a part (or a moment) in the whole. Each transcends itself to embrace all the rest, yet they are also related, as degrees in a scale, as the specific forms of the universal *Nous*.

The common-sense outlook (the "natural attitude") embraces the whole of experience at its peculiar level of reflection. There is no aspect of experience that is not subject also to scientific investigation. Morality and religion encompass the whole of life, including science as well as politics; and there is nothing that may not be expressed as, or embodied in, a work of art. The scope of philosophy is confessedly all-inclusive, the dialectician, as Plato claimed, being the spectator of all time and all existence. Yet each of these is but one specific aspect of intellectual activity, both practical and theoretical, and each includes and is included in all the others. Likewise, each is a whole in itself, differentiated into its own characteristic scale of forms.

Beginning with the life-world, we find it differentiating into practical and intellectual spheres, neither of them simple or uniform in character, yet each a whole composed of subordinate wholes. The practical life includes specific skills, social functions, moral duty, and political action, ranging recognizably over an ascending scale, each a subordinate self-contained whole, yet all mutually intertwined and overlapping. On the side of theory there are technology and science, art, philosophy, and religion, similarly interrelated. Again, each is a whole in its own right, each is specified in a gradation of forms and branches, and each interlaces with all the rest. It is not my present purpose to set any of this out in detail but to reserve such elaboration for later consideration. Our particular interest at the moment is the current world-concept projected by the sciences.

This world-concept is of a universe continuous and indivisible in space and time, deployed as a scale of comprehensive phases at successive levels: physical, chemical, biotic, sentient, and noetic, within each of which there is an analogous subordinate scale. Between them there are no breaks, none is independent of the others, but as they succeed one another so the prior are presupposed by and incorporated in those which supervene upon and grow out of them. Accordingly, the subsequent phases embody all their predecessors, and the latest sublates the whole prior scale, which in it is brought to a higher degree of actuality and self-sufficiency. The noetic level involves and reflects all the rest, for not only does it envisage and comprehend all the prior phases but also it is realized in the practical and intellectual activity of living beings, who are at the same time both organisms and physico-chemical systems, each drawing to a focus within itself ("inwardizing") its total environing world.

Here, then, the world comes to consciousness of itself and explicitly realizes its essential nature, in its reflective awareness and interpretive conceptualization by intelligent human beings. This is the basis and true justification of the Anthropic Principle, and from this vantage-point we may now return to it to reconsider its proper significance.

Notes

1. Errol E. Harris, *The Reality of Time* (State University of New York Press, Albany, NY, 1988), pp. 87–106 and 122–131.
2. Cf. Thomas Natsoulas, "Conscious Perception and the Paradox of Blindsight," in *Aspects of Consciousness*, ed. Geoffrey Underwood (Academic Press, London, 1982); G. G. Globus, "On 'I'; The conceptual foundations of responsibility," *American Journal of Psychiatry*, 137 (1980), pp. 417–422; Ludwig Wittgenstein, *Notebooks 1914/1916* (Blackwell, Oxford, 1961), and *Tractatus Logico-Philosophicus* (Routledge and Kegan Paul, London, 1961); Gilbert Ryle, *The Concept of Mind* (Hutchinson, London, 1949).
3. Natsoulas, op. cit., p. 82f.
4. Ibid., p. 85.
5. Cf. W. H. Thorpe, *Learning and Instinct in Animals* (Methuen, London, 1966).
6. Cf. Barrow and Tipler, *The Anthropic Cosmological Principle*, 7.1.
7. Cf. especially my "Is the Real Rational?" in *Contemporary American Philosophy*, ed. John E. Smith (George Allen and Unwin, London; Humanities Press, Atlantic Highlands, NJ, 1970).
8. Cf. E. Harrison, *Masks of the Universe* (Macmillan, New York and London, 1985).
9. Cf. my *Hypothesis and Perception*, Ch. VII; "Dialectic and Scientific Method," *Idealistic Studies*, III, no. 1 (1973); *The Reality of Time*, Ch. VII; "Contemporary Physics and Dialectical Holism," in *The World View of Contemporary Physics*, ed. R. F. Kitchener (State University of New York Press, Albany, NY, 1988).
10. Aristotle, *Politics*, I.2, 1252b25 (Jowett's translation).

11

The Anthropic Principle Revisited

THE SCIENTIFIC APPROACH

The physical cosmologist views the anthropic principle as a guiding thread to research. If what is observed is limited and shaped by the evolution and circumstances of the observer, outstanding problems may be soluble by taking this into consideration. For instance, life exists on a planet like the earth because there are heavy elements in the universe, of which the first to be produced is carbon, in the burning of red giant stars. The carbon nucleus is formed by combination of three helium nuclei, an event so improbable through chance collision as to be too rare to produce the required result. It could, however, occur in two stages: the combination of two helium nuclei (two protons and two neutrons) would produce a beryllium nucleus, which, combined with a third helium nucleus, would give carbon. But the isotope of beryllium, consisting of four protons and four neutrons, is wildly unstable (three other isotopes with more neutrons are in varying degrees much more stable). It was found, however, that resonance between the energy of the helium nucleus and that of beryllium vastly accelerated the formation of the latter, so as to counteract its instability. Still this was insufficient to account for the production of carbon in the required quantity. What was needed was a second resonance between carbon and beryllium, but neither the energy of the carbon nucleus nor any of its known overtones resonated with that of the other two. Human investigators, nevertheless, do exist to formulate this problem, life does exists on earth, carbon is available in the world, so there must be an energy level in the overtones of the carbon nucleus that resonates with beryllium. Fred Hoyle, arguing in this anthropic fashion, sought, and with the help of experimenters at the California Institute of Technology found the missing resonance.

Other arguments based on the WAP are less promising and lack conviction. It has been suggested, for instance, that the mass density of the universe, which determines whether its expansion will continue forever, or will slow down and eventually reverse to contraction, happens to be the critical amount that it is because, if it were not (as we saw earlier), either the distribution of matter would be too thin for galaxies to have formed or else the universe would long ago have collapsed in a big crunch. In either case we should not be here to observe it as it is. The fact that we are here, then, "explains" the critical density.

In this way, however, the remarkable seeming delicacy of precision in the initial conditions at the starting point of the universe, which determined this mass density, significant as it is, is hardly explained. The facts are merely acknowledged. If, however, the very nature of the universe were such that it could not be otherwise; and if from that nature the mass density followed necessarily, as also that intelligent life would eventually evolve, that would go a long way to explain why we are here to observe what we do.

Does such use of anthropic reasoning exhaust the significance of the principle, as enunciated by physicists? One may point to its strong version, which claims that the universe must have properties that permit life to develop at some stage of its history. But this, we have said, is no better than an obvious truism: because we are here to proclaim it, the universe must have such properties. A yet stronger version, now apparently supported by the most recent physical cosmologies, is less palpable, making the claim that no other universe would be possible. Even so, the various versions of the principle that physicists have offered fall short of the assertion that the very nature of the physical universe makes the emergence of life and the evolution of intelligent observers within it inevitable. Yet the scientific discovery that the physical world is an integrated, indivisible whole, and the philosophical analysis of the nature of wholeness, does give this result.

PHILOSOPHICAL REVIEW

Because the universe is a whole it must be a complete totality. Partial wholeness, as a final condition, is a contradiction in terms. What is partially whole is a part *of a whole*, and nothing can be partially whole that is not; but the whole of which it is a part cannot but be complete, and whatever is genuinely whole is perfect (in the original sense of *perfectum*, completely fashioned). This again is because a whole implies a structural principle that defines an integrated system and is fully expressed only in the ultimate totality. Being a principle of integral systematicity it is immanent in every part, resisting its partiality and driving it to augment itself, in order simply to maintain itself and so to develop beyond its provisional limits. The structural principle is, in consequence, necessarily and by its very nature dynamic.

Further, structure is a complex of relations that can be explicitly posited only in the comprehension of a conscious subject. It follows that, for reasons given above (Ch. 5), the physical world on its own has not the resources to fulfil its own holistic character. To do so, it must so complexify its physicality as to become a chemical and then an organic world, and that again, for similar reasons, must internalize its multiplicity in sentience and elaborate its intrinsic differences as a noetic world, in which the physical and the organic are sublated and re-presented in a theoretical system of unified science.

Consequently, the universe must be such as to provide, at some stage in its history, conditions for the emergence of intelligent life, and this must be what intelligent observers discover. Nor can there be, in this respect, any alternative, because any alternative universe, if it is a whole, must issue in a like result. Both the Weak and the Strong Anthropic Principle (WAP and SAP) are thus philosophically justified.

That the universe is *designed* to generate intelligent life has further implications which will be discussed in the next chapter. There is clearly strong evidence to support the idea, but the precise sense in which "design" is to be understood remains to be defined, and its implications both for humanity and for the universe at large need to be more fully explored.

The Participatory Anthropic Principle (PAP) is also justified by the above considerations, as has already been established. The way in which it has been couched states that observers are necessary to bring the universe into being. There may turn out to be a sense in which this is literally true—a sense with theological implications that cannot be appropriately investigated here and must be deferred for later discussion. But there is another sense that is relevant here in which the participation of the observer is essential to the realization of physical being. There is a wide range of degrees of being, and the physical universe may well *be* potentially (as Aristotle's prime matter was held to exist *in potentia*), while being fully actualized only when brought to self-awareness (as Aristotle's substance, *ousia*, reached full actuality, *energeia*, only in the last resort as *noeseos noesis*). For explicit realization, the potencies latent in the unperceived world must be brought to consciousness—they must be observed and known; and that can occur only if those potencies are nothing less than the potentiality of generating the very mind that observes them, so that in the observation the actual world is observing itself *in generationem*. "Study the distance to the stars and the structure of atomic nuclei . . ." writes Professor George Greenstein. "One piece of the universe is studying another. Through our existence the cosmos has become self-reflective."[1] That this should be so is required by the very nature of the universe as a systematic whole. The necessity for self-observation is, as it were, inscribed in its first beginnings and is implicit in its most fundamental physical laws.

The cosmic totality is unified diachronically as well as synchronically. It is the same whole in all its specific degrees and differentiations, physical, biotic, and mental. Accordingly, in our observation and in our interpretation of our own experience, what we discover is ourselves in the process of coming to be, and, concurrently, the world in the process of bringing itself to consciousness in our minds. The two processes are one and the same, and the outcome is the actualization at once of both subject and object. Aristotle had this right more than two thousand years ago when he maintained that in sensation the faculty (or disposition) and the object were both actualized in one.[2] Accordingly, intelligent observers are necessary to bring the universe into being, if that means realizing in full the implications and potencies of its essential nature.

The Final Anthropic Principle (FAP) affirms the necessity that intelligent information-processing should come into existence in the universe and then continue indefinitely. Contemporary scientists make this conditional upon the sufficient development of artificial intelligence (von Neumann probes). That, however, is a contingent development which could contribute to the persistence of information-processing but would hardly guarantee what is really essential, namely, reflective consciousness.

The hypothesis is that information-processing, once it has reached a sufficient degree of sophistication, will be able to reproduce itself and to proliferate, will become capable of space exploration and colonization, will be invulnerable to destructive conditions that mere flesh could not tolerate, and so, eventually will overrun the entire universe.[3] But this hypothesis is seriously flawed. Physicists seem too ready to equate artificial intelligence with rational thinking. The former is the manipulation of "bits" of information by mechanical means; the latter is self-conscious, self-reflective reason. It is by no means established that the two are indistinguishable, much less identical. Self-conscious thinking, we have maintained, is self-transcendent. It is far from clear that information processing has any such propensity.

The test for artificial intelligence that Turing prescribed was that its answers to questions should be indistinguishable by an impartial observer from those of a human respondent. Such a test may be passed by a well-programmed machine. But how can we decide whether such a machine is capable of reflective self-criticism? Suppose you put the question: "Shall I compare thee to a summer's day?" What answer would you expect? A human examinee might take the question as a request to continue the quotation. Or the question might be regarded as rhetorical or as a serious request for information or advice. A machine programmed to answer appropriately in any of these ways could not, without further information, be distinguished from a living person. But how can we test its critical ability? If asked whether the answer given was the right one, the machine should reply in the affirmative so long as it has not malfunctioned and can match its response with its programme. A human respondent will consider a different criterion; but what could distinguish one from the other? If there is no way of knowing whether artificial intelligence is self-reflective and thus self-transcendent (which almost certainly it cannot be), whether the critical threshold of consciousness will have been crossed will remain undisclosed.

Again, a computer can (at all events, in principle) be programmed to construct another machine like itself. It can be programmed to write its own programme or any other. It can be programmed to detect errors and correct them, or to register failure and provide for its remedy. None of this necessarily involves self-transcendence. Not even the so-called memory of the machine is memory in the true sense. It is, at best, storage and retrieval. When I misremember I can be, and often am, aware of the fact. Could a machine be said to misremember or to be able to recognize its own error? Could it be programmed to recognize its own malfunctioning?

It has been argued (we recorded earlier) that time enough has elapsed since the Big Bang for intelligent life to have appeared on other planets attending other stars than the sun, and for it to have developed far enough to invent von Neumann probes and to send them out on the conquest of space, so that by now they should have reached the solar system. No reliable evidence that they have done so is at hand. Therefore, the argument concludes, there is no extra-terrestrial life. Even if there is, the evidence indicates that it has not evolved to the required extent. The viability of the Final Anthropic Principle, then, depends entirely on the human race, now heading for environmental disaster that threatens its extinction before von Neumann probes emerge from the realm of calculated hypothesis. That makes FAP predictions precarious in the extreme.

Of whatever von Neumann probes may be capable, if they are not self-reflective they could hardly conceive of the universe of which they are members, and could thus not bring to fruition the whole, which for its completion demands self-consciousness. If it transpires that they operate without consciousness, they can never finally effect the collapse of any probability amplitude represented by a Psi-function. For the FAP to hold, these demands must be met in full, but it is far from clear that von Neumann probes could meet them. If Roger Penrose's argument (to which reference was made in Ch. 8, above) is to be credited, a computerized automaton which does no more than operate an algorithm (however complex or sophisticated) cannot generate insight or develop consciousness.

And nothing less than self-reflective, insightful cognizance is necessary to bring the systematic structure of the whole to full fruition and explicit evidence. No reliance therefore, can be placed on the development of mechanical information-processing to give validity to the FAP. Nevertheless, there may be other reasons for accepting the Final Anthropic Principle that should not be left out of account.

If the universe as a systematic whole can only achieve completion in reflective consciousness, so that such self-awareness is, as it were, built into its implicate order, the emergence of intelligent life must be intrinsic to its very nature. In that case, the extinction of humanity in this planetary system will simply stimulate the development of intelligent creatures in some other, as the plucking of dead heads from a rose bush encourages the blossoming of new roses. If this is so, the Anthropic Principle is final, and intelligent mentality can never be eliminated from the universe.

As matters now stand, at least preliminary conditions for the fulfilment of the potentialities inherent in the universe have been met by the development of human intelligence, but they are not yet satisfied in full. The human mind works subject to the limitations of the animal organism and is still finite, despite its premonitions of infinity. Human science, for all its achievements, is far from complete, is at best an approximation, is riddled with hiatuses that scientists are for ever striving to fill in, and puzzling contradictions that they are always striving to resolve. Human knowledge is an aspiration rather than a final attainment, and a persistent endeavour to discover what lies beyond its present comprehension. It seeks to understand the universe in which it finds itself and the true nature of which it senses in dim forecast. This premonition is possible because of the immanence in the human thinker of the principle of order expressing itself throughout the entire gamut of natural forms; and although that principle comes to consciousness in human experience, its full expression does not culminate in the exact and the empirical sciences.

Science labours under self-imposed limitations. Its theories are essentially hypotheses and do not profess to state the truth categorically. It makes initial presuppositions that remain unexamined unless and until the current "paradigm" breaks down, and for the most part the scientist is unaware even of making them. Reflection has to return upon itself yet again to discover these limitations and to try to transcend them at a further level of syngnosis.

Science has to go over into philosophy, the purview of which includes morality, art, and religion as well as knowledge. These are then viewed not only from without, as science itself is capable of viewing them, but, at the same time, from their own perspectives, while the endeavour is made to interrelate them all in a single coherent vision. The noetic scale has to ascend to a metaphysic that sees things together (as Plato averred), and is, moreover, also a theology (as Aristotle was aware).

Hitherto we have approached this aim through a review of scientific theories because the Anthropic Principle has been devised by scientists. For its attainment, however, the noösphere must be examined more completely and its structure followed in more intimate detail. But before embarking upon this task, we need to examine one more consequence of anthropic reasoning that has attracted the attention of some physicists. We have yet to examine the basis, if any, of what I have dubbed the Teleological Anthropic Principle. This question arises from the scientific outlook even though it goes (by implication) far beyond it, so its consideration justifies reserving it for a separate chapter. The implications that go beyond science would warrant its deferral even further, until after the other aspects and elements of the noösphere have been examined in detail, but some anticipation is permissible in order to complete the examination of the Anthropic Principle in its scientific context.

Notes

1. George Greenstein, *The Symbiotic Universe*, p. 48.
2. Cf. *De Anima*, II. xii, 424a17–19 and III, ii. 425b26.
3. Cf. Barrow and Tipler, *The Anthropic Cosmological Principle*, Ch. 10, 6.

12
Argument from Design

GOD AND THE NEW PHYSICS

The enunciation by contemporary physicists of the Anthropic Cosmological Principle has stimulated interest in and discussion, hitherto rare among physicists, of the question whether recent developments in physical science have provided new evidence for the existence of God, and has revived in particular the traditional argument from design, which points to the intricate and seemingly teleological arrangement of physical and biological entities as evidence for the existence of an intelligent creator and designer. Part of the adverse criticism that this argument has aroused in the past has been uncertainty about the extent of such evidence, the expression (for instance, by Hume[1]) of doubt that we can discover in nature clear signs of orderly structure, such as is commonly alleged by exponents of what is sometimes called the Physico-theological Proof of the existence of God.[2] The contemporary revival of the argument, therefore, must direct attention to the question whether the new physics has, in fact, discovered "design" in the physical world.

In any consideration of this question, much will depend on the meaning we attach to the word "design" and how, in consequence, we interpret the argument based upon it. These issues will be my first concern in this chapter.

Argument from design has a very long history that has been traced back by Barrow and Tipler[3] to the ancient Greeks in the West and even further in Eastern thought. But it has not always had the same significance nor been applied to the same objective. Sometimes it has been used to support the belief that nature in general is designed for the benefit of mankind; at others, especially in Western theology and philosophy, as well as carrying this implication it has been deployed to establish the existence of a divine creator of the world. My object here will be to show that neither of these uses is legitimate unless carefully qualified, but also that, if properly understood, both can, in an important sense, be vindicated.

Meaning of "Design"

In discussion of this type of argument (certainly by Barrow and Tipler) there has been much confusion, or at least not much clear distinction, between notions of design, purpose, teleology, and final causation; and, while all these concepts are relevant to the argument, they are by no means all synonymous. Distinctions between them are not unimportant, and clear understanding is unlikely to be reached unless the various nuances are respected.

The word "design" is itself ambiguous, having at least eight different meanings listed in the *Oxford English Dictionary*. They include "a plan or scheme conceived in the mind and intended for subsequent execution"; "purpose, aim, intention," and "the thing aimed at, the end in view." It also means (again quoting the *O.E.D.*) "contrivance in accordance with a preconceived plan; adaptation of means to ends; prearranged purpose." It has the further sense of "a plan in art," or "preliminary sketch"; "a plan for a building," as well as "the combination of artistic details or architectural features which go to make up a picture, statue, building, etc."

Several of these definitions emphasize the schematic character of design. It is a plan or scheme, a combination of artistic details, contrivance, a preliminary sketch (or, a sort of blueprint), and it is pre-arranged. In other words, it is some kind of structural form. Other definitions stress the idea of purpose: it is conceived by the mind, intended for subsequent execution; it is a purpose, aim, intention. Again, it is the object of such intention and purpose, the thing aimed at, the end in view. In short, there are three main meanings, distinct though not unconnected: (i) a pattern or blueprint, a scheme or systematic arrangement; (ii) a purpose, aim, or deliberate intention; and (iii) the end aimed at, the object of such intention or purpose. The traditional arguments from design use the word in all of these meanings, without clear distinction. As stressing the notions of end or goal, as well as of purpose or deliberate intention, it has been called the Teleological Argument, but in many instances the idea of pattern or structure is more prominent. In recent discussions a new name has been proposed for the second type: the Eutaxiological Argument. But few if any proponents of the argument recognize this difference, which Barrow and Tipler, however, have pointed out (loc. cit.). If the first sense is implied, then an intelligent mind is presupposed forming the deliberate intention and conceiving the purpose. If the second, an intelligent designer or artificer is postulated.

The concept of purpose always involves conscious, deliberate intention pursuing an end or goal, but while this is included in what is recognized as teleological, because that word etymologically implies some action aimed at or directed towards an end or final objective, teleology has a wider connotation. Any process that is directed towards a final state or condition, such as the ontogenetic development of an embryo or the growth of a plant, may be regarded as teleological without presuming any conscious purpose or deliberate intent. Final causation is held to apply in both examples, and is the idea of a cause by which that process is modified, directed, or regulated to produce the ultimate outcome, as opposed to an efficient cause which determines the effect in accordance with its own prior nature.

Design, if meant to imply purposed plan of action, obviously involves final causation in the above sense; but if it means no more than ordered pattern, this is not necessarily the case. The pattern of a snowflake, for instance, or a section through an onyx stone may reveal a design that has involved no planned or purposed designing. Nevertheless, we shall presently find that all these terms and their various meanings are connected and are subject to a common condition.

Objections have been raised against the idea of final causation (by Francis Bacon, for instance), on the grounds that it can explain nothing and that appeal to it is no more than the reiteration of what is already known (as in Molière's satirical declaration that opium causes sleep because of its soporific tendency). Further, disapproval is expressed of the assumption in final causation that a future event can influence a prior process. Although these objections have some force in certain cases, they certainly do not always apply, depending on the way in which teleology is to be understood.

When we ask if physicists have discovered evidence of design in the universe, it is important to observe the difference between the two chief possible meanings of the word "design," because if taken in the sense of purpose or deliberate intention, the answer is very likely (if not certainly) to be in the negative, while if understood in the sense of pattern or structure an affirmative answer may well be forthcoming. But if the two meanings are confused, evidence in favour of one may be mistaken for evidence in favour of the other. At the same time, if we find that the two senses have a common basis, the discovery of pattern and structure may not be irrelevant to the presumption of teleology and final cause.

UNSATISFACTORY FORMS OF THE ARGUMENT

The most naive form of the argument from design sees the products of nature as intended and created purely for the benefit of human beings. This is the corollary of an extreme anthropocentrism for which there is no scientific or rational justification, even if we do not go to the ridiculous extremes of those who have contended that cork trees were created by God to provide men with stoppers for wine bottles, or the ultimate inanity of Bernadine de St. Pierre's suggestion that dogs are commonly of two colours, light and dark, in order that they may be distinguished from the household furniture.[4]

Less ludicrous examples of this argument are still unacceptable, if only because there is at least as much evidence against them as could possibly be mustered in their favour. A Chinese story from the 3rd century B.C. depicts the host at an ancestral banquet discoursing on the generosity of Heaven in providing grain, fish, and fowl as food for the benefit of mankind. He is then refuted by the response of a small boy who points out that Heaven has also created mosquitoes, tigers, and wolves that feed on men.[5]

The argument in somewhat less unreasonable guise has frequently been advanced in more modern times, and as frequently dismissed, if only for its unwarranted implication that so comparatively minute a celestial body as the Earth, and, even more, such insignificant and defective creatures as humans, might be the end and purpose for which the vast expanses of the universe and its contents were intended as means. This objection was voiced as long ago as the 6th century A.D. by Boethius, who writes:

> Thou hast learnt from astronomical proofs that the whole earth compared with the universe is no greater than a point, that is, compared with the sphere of the heavens, it may be thought of as having no size at all. Then, of this tiny corner, it is only one quarter that, according to Ptolemy, is habitable by living things. Take away from this quarter the seas, marshes and other desert places, and the space left for man hardly even deserves the name of infinitesimal.[6]

How could so minute a creature be the end and purpose of the entire and vasty cosmos? The same demurrer has been repeated from time to time up to the present day.

The implication of anthropocentricity, however, may be defended on other grounds, and may not be summarily dismissed for its seemingly crude teleology. Although it may not be allowed to support naïve conclusions, there will be occasion in the sequel to return to it.

THE SUPREME ARCHITECT

In the 17th and 18th centuries, after the Copernican revolution had introduced the mechanistic conception of the world that was developed by Newtonian science, the argument from design, based on this view of the universe as a machine, was used as a proof of the existence of God. It was by no means new, but it now took on a new look and was widely espoused by such prominent scientific figures as Robert Boyle, John Ray, William Harvey, and by Newton himself, who wrote in the concluding Scholium of the *Principia*:

This most elegant system of suns, planets and comets could only arise from the purpose and sovereignty of an intelligent and mighty being . . . He rules them all, not as a soul but as a sovereign lord of all things, and because of His sovereignty He is commonly called "Lord God almighty."

At the same time, the argument was rejected with considerable force and on formidable grounds by Spinoza and Hume, as well as by many others. Perhaps its most persuasive statement was made by William Paley,[7] and its most influential refutation was given by Kant.[8]

The reasoning was that a machine can be constructed only by an intelligent mechanic and cannot be assembled by chance concatenation; and as the universe had been shown to work like a machine, and bodies to move according to orderly mechanical laws, there must be a supreme artificer who had created it and maintained its operation. Paley used the example of a watch, of which the intricate contrivance of the interrelated working parts could be explained only by assuming that it had been made by a skilled and intelligent watchmaker. On this analogy, the observed celestial clockwork must be the contrivance of a Supreme Technician or Architect.

This form of the argument is eutaxiological, stressing the meaning of design as structure rather than as deliberate intention, but as a machine is a device normally constructed to serve some human purpose, it seemed natural to assume that the world had been designed with an analogous end in view. Thus the teleological aspect was not excluded.

It must be borne in mind that at this period, and for centuries prior to it, the prevailing belief was in the existence of an omnipotent God, the creator of Heaven and Earth, so that the argument from design was simply an additional support for a generally accepted belief, mitigating any possible doubt on supposedly scientific or rational grounds. The counter-arguments of such thinkers as Hume and Kant were not so much denials of the existence of structure or order in the world as submissions that such evidence of order and conformity to law as had been found was not sufficient to warrant the conclusion to a creator God who must exist of necessity in His own right; whereas the existence of an artificer was itself a contingent fact that required further explanation.

The mechanical conception of the world, however, was built on the presupposition of the existence of particles, or mass-points, in external relation one to another, moving under the influence of imposed forces proportional to the inverse square of the distance. The forces were not dependent upon any intrinsic characters of the particles, which were not presumed to have qualities other than mass, so that their mutual disposition could well be the result of mere chance and was initially presumed to be chaotic. To produce order in this chaos, it seemed necessary to postulate the intervention of an intelligent designer. But, in fact, a random shuffling of such particles over a sufficiently long period of time (and the available time was usually assumed to be infinite) could produce all possible configurations. Consequently, inference from apparent order to a Supreme Architect was unwarranted, at least without further, more compelling, evidence.

Yet more serious was the failure of the mechanistic picture of the physical world to provide any account of the occurrence within it of observers conscious of its structure and working. Attempts made by thinkers such as Hobbes and Locke to explain knowledge of the world as derived from the mechanical effects of external objects upon the human sense organs broke down utterly, to result in the scepticism of Hume. Kant sought to remedy this debacle by making objects conform to concepts, instead of deriving "ideas" from external causes, so that the objectivity of scientific knowledge could be preserved by explaining it in terms of the inherent tendency of the mind to organize and systematize its otherwise confused sensuous experience. But then the objects of science had to be recognized as purely phenomenal, only appearances to us, and the categories determining their order and arrangement could not be applied to "things-in-themselves." In that case, no inference was legitimate, from the discovery of order and pattern in the physical world to any Divine Architect, except so far as He might be held responsible for the existence and nature of the human mind.

More compelling evidence for design seemed, in due course, to come from the discoveries of biology. Even Kant was impressed by this and maintained that biological facts could not be made fully intelligible solely by mechanical laws but required the idea of teleology for their interpretation. The appeal to physiological and biological phenomena thus became more common in versions of the argument from design. But, despite his sympathy and preference for the Physico-theological argument (as he called it), Kant's objections to it as a proof of God's existence were not overcome by biological evidence. Teleology, he held, was at best a "regulative idea," a mere postulate that did not constitute objective knowledge even of phenomena, much less of things-in-themselves. In the case of phenomena, moulded by the schematized categories, the only objective knowledge was of mechanical laws, which had already failed to afford evidence for God's existence. So, for Kant the Teleological Proof remained invalid.

But what seemed finally to dispose of the design argument was the rise of Darwinism in the middle of the 19th century, which, in effect, established the possibility of seemingly teleogical order coming into being by chance variation and natural selection (itself a purely fortuitous process). The purposive aspect of design thus became an entirely unnecessary and unwarranted assumption, and no legitimate conclusion about an intelligent cause could be drawn from the observed adaptation of living forms to their environment or the suitability of their construction and behaviour to their biological needs.

It is not my intention, however, to discuss in further detail either the argument itself in the form so far presented or the various refutations and objections opposed to it. They were all predicated on the mechanistic conception of the universe presupposed by Newtonian science, and could not be justified on that presumption. The laws of formal logic consonant with the empiricist metaphysics encouraged by Newtonian physics rendered the argument from design (along with the other traditional proofs of God's existence) invalid. But the mechanistic conception of the world has now long been abandoned, at least by physicists, and we must therefore reconsider the argument in the light of the new physics of the 20th century, which demands an entirely different approach.

THE CONTEMPORARY WORLD-VIEW

In earlier chapters I have been at pains to show that the most significant revolutionary effect of the physics of relativity and quantum theory has been to generate a new view of the physical universe as a single, indivisible, undissectable whole, in which phenomena and events are necessarily interdetermining. It is a single system governed by a unitary dynamic principle, of which the mathematical expression has not yet been discovered but which is confidently foreshadowed in the most advanced theories. Contemporary physics offers a wealth of evidence to support this conception, which is augmented by the results of research into complex dynamic systems, as well as by the findings of the sciences of biology and ecology.

As I have repeatedly argued, a whole such as this, or, for that matter, any whole, properly so-called, cannot be a blank, undifferentiated unity, but its integrity must depend on the interconnection of parts internally related one to another, in accordance with an organizing principle. In the dimension of time, this principle is dynamic, generating a graded scale of subordinate wholes in which it is specified. Whatever the detailed interrelation of elements may be, or the particular structure of specialized areas, it always resolves itself in this way into a progressive series of dialectically related forms each of which is a specific exemplification of the universal dynamic principle governing the whole. The forms are ranged in a series of increasing degrees of adequacy to the nature of that organizing principle, progressively revealing more fully its universal character.

Now, a design, in the sense of pattern or structure, is obviously such a whole. Its parts and elements must be interrelated systematically according to some principle of order and arrangement. They will, that is, be internally related, and so will determine one another in accordance with the over-riding principle of structure. Every design must be a whole, for it can be recognized as a pattern only by reference to its completed composition. A fragmentary fossil, for example, can be recognized as the remains of a living animal only by envisaging the complete skeleton, and once the principle of structure is known, the paleontologist can reconstruct the entire anatomy from the broken remnants.

It is the immanence of the principle of order in the parts of a structured whole that constitutes its teleology (which, despite etymology, is not confined to goal-directed behaviour). The nature, disposition, and be-haviour of the part can be accounted for only by reference to this immanent principle, where the discovery of efficient causes leaves unexplained the structure and dynamic regulation of movements within the whole. Such explanation of the part in terms of the whole is what is properly termed teleological. In the case of a dynamic movement or a genetic process, subjection of the phases to the governance of the principle of wholeness will determine the end, which is typically the completion of some whole; thus a teleological process is one of genesis of a whole, and if the process is consciously directed it is purposive.

The objections to teleological explanation so frequently voiced since the 16th-century reaction against Aristotle now fall away. The complaint that it reverses the order of causation, making a future event the cause of the present no longer applies, because the teleological effect is now seen to result from the principle of systematic wholeness immanent in the part and in the process of generation of the total structure. Nor is there necessarily any presumption of conscious intent or deliberate purpose, although that is a highly developed instance of teleological action; not, however, simply because it is directed towards an end but because it consciously seeks to complete a planned structure or design. The goal of purposive activity is not merely, and often (especially when the endeavour fails) not at all, its terminal stage. The aim of a musician composing a symphony is not the final chord but the symphonic whole. The goal of the athlete is not simply to break the tape but to complete the course in the shortest time. As Aristotle rightly perceived, "not that which is last deserves the name of end, but that which is most perfect."[9]

In the contemporary view of the universe, design (regarded as structure), teleological process, and purposive action are not merely alternative senses of the word; they are themselves phases in a scale of forms exemplifying in differing degrees a universal concept. And they correspond to a parallel ontological scale, from space-time to metabolism, and from physiological process to conscious behaviour, specifying a universal principle of organi-zation.

Design, then, may be taken to cover all these: it may be a blueprint, an anatomical structure, a mechanical device, a plan of action, or an aesthetic composition. It may also be a genetic process or an evisaged goal. In every case, the essential principle is holism, organization, or structure, and its teleological character is inescapable.

In the order of nature the second of these scales is continuous. Intelligent beings, capable of observing their surrounding world and reflecting upon that experience, have evolved from more primitive organisms that have emerged from a primaeval "soup" of chemically reacting physical substances. Geometrical and physical design, disclosed by contemporary physics, issues in organic design, which again generates purposive design.

But none of this is any more than implicit until it is grasped and comprehended in conscious experience, in observation and interpretion, by the scientist and the philosopher. A design is a relational complex, and no relation resides in any one term but is a whole involving two terms or more. It can be comprehended and realized as a whole, only as cognized by a subject who transcends the singular parts, and is not confined to any of them. When relations are internal, as they are in every design, while the whole governs the nature of and relations between its parts, the parts (the terms in relation) are differently affected according to the level they occupy in the dialectical scale. Physical entities move and interchange as determined by the energy system, but in so doing they merely react to the other elements in the system in consequence of the holistic influence. In the organism there is sentience, enabling it to respond to felt conditions, but it still does so blindly or by instinct. Only at the conscious level is the relation between the subject and its surroundings explicit, and only as cognized is the whole fully realized.

The whole gamut of natural forms, moreover, from the physical to the self-consciously reflective, constitutes one continuous whole of interrelated phases. It is a whole of wholes, a system of systems, a design comprehending a whole series of subordinate designs. And this design is in principle complete *ab initio*. So its self-awareness at the conscious level is already implicit in the initial phases. But while the principle of order is operative from the start, it can only be fully actualized when explicated in consciousness. This is necessary to its holistic self-actualization, and its *nisus* to self-explication is the ground-spring of its self-unfolding. Hence, intelligent comprehension of itself is what the natural order itself generates.

The process at one and the same time produces and is contained in the outcome; the prior phases of the scale are embodied in their subsequent product. Process and end overlap and each is necessary to the other. The design in nature is not fully realized, not completely explicated, until consciously observed and interpreted. It becomes fully actual only in being systematically cognized and intellectually expounded, but this can occur only when intelligent life has evolved in organisms emergent from the chemical and physical matrix.

The whole which contemporary physics has revealed, therefore, necessarily involves the generation of its own observation by intelligent beings, in whose minds it brings itself to consciousness. This is the true significance of the Anthropic Cosmological Principle, and is the modern version of the contention, derived from Aristotle (whose argument is, in principle, the same as that presented here[10]), that natural forms exist for the sake of humankind. It is the proper legitimization of the first form of the argument from design, and it involves none of the stupidities, nor the logical defects, of its earlier expressions. The question arises whether it is possible to go further and to vindicate the second.

THE EXISTENCE OF GOD

A whole or design, I have maintained, cannot ultimately be fragmentary. It must in principle be complete. The whole of nature that constitutes the cosmic design deploys itself as a scale of forms, which (it was shown in Ch. 2) cannot go on forever but must ultimately culminate in a completed totality. Nor can its self-manifestation be only partial. Were the scale endless, it would never be complete. The universal, which its phases instantiate, would never be realized and the principle of order would dissipate. But, in that case, there would be nothing to instantiate and there could be no instantiation and no scale. In principle and in fact, there must be a culmination of the scale that is both final phase and all-encompassing—an absolute, actual whole, totally self-contained and self-sufficient, and completely realized. It must sublate in itself the entire process of its self-specification, so that end and process overlap.

It would be a mistake to imagine that this culmination can, or needs to, appear in time, for it must encompass all time in itself while nevertheless enduring throughout time. It does so in the same way as human consciousness transcends the present and includes at once both past and future while it continues to endure and participate in the flux. The culminating phase of the scale does this likewise, for it is the fulfilment of the organizing principle universal to every phase and every existent. It is immanent throughout all process, for every process is a manifestation of its self-differentiation, contributing at its specific level and in its peculiar degree to the final consummation.

In some measure, the human intellect does satisfy the above description, and the human mind is a relatively late phase in the scale. It does sum up and re-present the entire prior process of its own development. Yet it falls far short in many respects of the final consummation. Human science and philosophy taken together and *en masse* are immeasurably less than omniscience. Science is constantly advancing, current theories are continually proving unsatisfactory and are being revised and improved. In short, human knowledge is notoriously incomplete. The deficiencies of human conduct are so gross that little time need be lost in compiling a catalogue. It follows, therefore, that although the *scala naturae* requires a conscious mind to make it explicit, and although the development of intelligent life in human personality conforms to this requirement, it is far from fulfilling it completely and is at most an adumbration of the ultimate totality. Accordingly, if the whole must be complete in principle, we must extrapolate further.

Design resolves itself into a dialectical scale of forms, of which each is a provisional whole exemplifying in its specific degree the universal principle of organization. Of such a scale three characteristics need emphasis: (i) it is in principle absolutely complete; (ii) its completeness involves total explication in absolute self-consciousness; (iii) the final phase, like all others, must transcend and at the same time include and comprehend all its predecessors—that is, the final phase both must be, and yet must transcend, the sum-total of all the parts. It must be such as no conception or existence can exceed. Consequently, it answers to the definition of God derived from Anselm and Aquinas, and formulated by Descartes and Spinoza, as a perfect being (that is, a being who is totally complete, totally self-sufficient and self-sustaining—*perfectum*). That has already been shown to involve self-consciousness and, sublating all subordinate phases, congnizance of the entire cosmic totality. It will be "that than which a greater is inconceivable."

As totally explicit in transparent self-consciousness, this consummation of the cosmic scale is an omniscient mind. Being the actualization in full of a universal principle of order, which is immanent in every phase and every detail of the scale, it is the Alpha and the Omega of all being. Because the universal principle is immanent in every part, it is what generates and determines the nature of every entity, and its activity is nothing more nor less than its own self-differentiation in and as the spatio-temporal world. But its ultimate realization is a transcendent comprehension and self-concious realization of the whole. It is thus all-creative and all-powerful, as well as all-knowing and absolutely self-complete. All this is necessarily entailed by the very concept of design.

If God is conceived as the absolute universal principle of order manifesting itself in and as the universe, and transcending all finite phases, the argument from design, as a proof of his existence, can be justified in this, its modern form, without requiring any inference from a contrived plan to a Supreme Architect (unless these phrases are used metaphorically). The analogy to the human artificer remains inept and illegitimate because God does not construct out of an alien material according to a preconceived blueprint. His knowing and conceiving are immediately and simultaneously his self-manifestation in and as the world—his creative power. Thus all natural activity is the process of God's creation, and all nature is God's self-revelation. Nor is this Pantheism, because God is not only immanent in nature but also transcendent beyond the finite. On the other hand, human construction is in some sense the image and premonition of God's creation, insofar as it is the expression and objectification of an experience that has been generated in the course of the self-realization of the universal principle immanent as well in human personality as in every natural form.

In this contemporary version of the argument, the conclusion emerges from the necessary dialectical implication of the very concept of design. There is no need to seek an explanation outside the structure and processes of the natural world for what occurs within it. Professor Paul Davies, in his admirable book *God and the New Physics*, repeatedly raises the question whether we are obliged to assume the existence of a supernatural creator to account for certain remarkable features of the physical universe, such as large cosmological number coincidences or the extraordinarily delicate balance of primordial forces immediately subsequent to the Big Bang, without which the detailed structure of the present universe could not be what it is. He gives no categorical answer to the question pro or con, but speculates whether, perhaps, as he puts it, some perfectly natural explanation might not eventually be discovered.[11]

The right answer would seem to be that whatever legitimate explanation were given would be a natural explanation, and the obligation to assert the existence of God is not the failure otherwise to account for surprising coincidences. It arises directly from the inevitable implication of the unity of the universe as a necessarily self-differentiating and self-explicating whole that cannot but be complete as such, and unremittingly requires a consummation in a totally self-sufficient, self-sustaining, and self-conscious mind.

This conclusion has the rare advantage that it is not a resort to God as a cloak to cover our ignorance, but is the logical consequence of the very nature of our knowledge and of the structure of the universe as discovered by empirical science. Here, then, is the argument from design in modern dress, and the epithet Physico-theological describes it nicely. It is the argument for the existence of God that follows most readily from the discoveries of the empirical sciences, and I have discussed it here because it has been publicly and deliberately resurrected by the physicists themselves. Empirical science, however, is not the eternal truth—or at least not all of it. It has been admitted that conceptual schemes contain frustrating inconsistencies and from time to time require adjustment and modification, periodically amounting to revolutionary change. Even now there is an alternative theory of the physical world waiting in the wings to replace the belief in the Big Bang, with all its consequences and corollaries.

Dr. Hannes Alfven has developed a theory of plasmas that can account for most of the astronomical phenomena while avoiding one of the most serious difficulties of the current theory.[12] The view most widely held at present implies a quantity of matter in the universe far in excess of what can be observed. To supply the deficiency, the doctrine of "dark matter" has been conceived, with the grave disadvantage of assuming that the greater part of the physical universe is unobservable. Dr. Alfven's theory does not require this assumption and seems in many other ways to agree with the observable facts.

Few physicists seem ready, as yet, to follow Dr. Alfven's proposals, but a shift to his alternative theory would involve many fundamental changes in the accepted world-picture. It would, for example, substitute electromagnetism for gravity as the dominant force in the space-time whole, and it would abolish black holes. But the one relevant consideration for the argument from design, as here presented, would be whether the universe were still conceived as one unitary system. If it were, and as long as the dialectical series of ascending forms remains unassailed, that argument would stand.

Further, science is but one facet of a wider and more complex noösphere. It is inseparable from society and all that entails, and the social sciences have not yet, in the above exposition, been consulted. In conjunction with them, the philosophical system I have sought to outline holds further and even deeper implications, which it is important to pursue. These, however, I propose to hold over as the subject of another volume.

Notes

1. David Hume, *Dialogues concerning Natural Religion*, Part II.
2. A term introduced by William Derham (1657–1735) in his Boyle Lectures for 1711–12, and later adopted by Kant.
3. Cf. Barrow and Tipler, *The Anthropic Cosmological Principle*, Ch. 2.
4. Ibid., p. 92.
5. M. Eliade, *From Primitives to Zen: A Thematic Source Book of the History of Religions* (Harper and Row, New York, 1967), p. 94, quoted by Barrow and Tipler, *The Anthropic Cosmological Principle*, p. 95.
6. Boethius, *The Consolations of Philosophy*, II, Prosa vii.
7. Cf. W. Paley, *Natural Theology* (New York, 1824).
8. Cf. I. Kant *Kritik der reinen Vernunft*, *Transcendentale Dialektik*, A620, B648– A630, B658.
9. Aristotle, *Physics*, II, 2, 194a30.
10. Aristotle represents the world as a *scala naturae* in terms of matter and form. Matter is potentiality and form actualization, so the matter is said to be "for the

sake of" the form. Each stage is constituted by the imposition of new form on the prior stage as proximate matter. The human soul is the form of the body, which is living matter; life is the form imposed on the *homoiomerē* (mixtures, akin to chemical compounds), which again is the form imposed on the elements (earth, water, air, and fire), which are forms imposed on differential combinations of the opposites (hot, cold, moist, and dry). These again inform primary matter (pure potentiality). At each stage the matter is "for the sake of" the form. So all the lower forms are for the sake of the human soul, which itself exists "for the sake of" its ultimate good, the activity of reason, in imitation of God's activity, the form of all forms.

11. Cf., Paul Davies, *God and the New Physics* (London, 1983; Penguin Books, Harmondsworth, 1984, 1986). Professor Davies does suggest, however, in the context of the hypothesis, that there might be only one possible sort of creation: "God is the supreme holistic concept . . ." This, he says (p. 223), would be "a natural, as opposed to a supernatural God."

12. See *New York Times*, Feb. 28, 1989.

Bibliography

Alexander, S. *Space, Time and Deity*. Macmillan, London, 1929.
———. *Spinoza and Time*. G. Allen and Unwin, London, 1921.
Allport, Floyd. H. *Theories of Perception and the Concept of Structure*. Wiley, New York, 1955.
Ames, Adelbert. "Visual Perception and the Rotating Trapezoid Window," *Psychological Monographs*, 63, no. 7 (1961).
———. "Reconsideration of the Origin and Nature of Perception," in S. Katner, *Vision and Action*. Rutgers University Press, New Brunswick, NJ, 1953.
Aristotle. *De Anima*, trans. Hugh Lawson-Tancred. Penguin Books, Harmondsworth, 1988.
———. *Metaphysics*, trans. W. D. Ross. Clarendon Press, Oxford, 1924.
———. *Physics*, trans. Edward Hassey. The Clarendon Press, Oxford; Oxford University Press, New York, 1983; trans. P. H. Wicksteed and F. Cornford. Heinemann, London, 1929, 1934.
———. *Politics*, trans. E. Barker. Clarendon Press, Oxford, 1946; trans. T. A. Sinclair. Penguin Books, Harmondsworth, 1981.
———. *Basic Works*, ed. R. McKeon. Random House, New York, 1941.
———. *Works*, ed. W. D. Ross. Clarendon Press, Oxford, 1928, 1963.
Armstrong, D. M. *Perception and the Physical World*. Routledge and Kegan Paul, London; Humanities Press, Atlantic Highlands, NJ, 1961.
Austin, John. *Sense and Sensibilia*. Clarendon Press, Oxford, 1964.
Barnes, Winston H. "The Myth of Sense-Data," *Proceedings of the Aristotelian Society*, XLV (1944–45).
Barrow, J. D., and Tipler, F. J. *The Anthropic Cosmological Principle*. Oxford University Press, Oxford and New York, 1986, 1988.
Bartlett, Sir Frederick. *Remembering*. Cambridge University Press, Cambridge, 1961.
Bastin, T. (ed.). *Quantum Theory and Beyond*. Cambridge University Press, Cambridge, 1971.
Bates, M. *The Forest and the Sea*. Random House, New York, 1960; Vintage Press, New York, 1965.
Bergson, Henri. *l'Èvolution Creatrice*. F. Alcan, Paris, 1907; A. Skira, Geneva, 1945.
———. *Matière et Memoire*. F. Alcan, Paris, 1896; A. Skira, Geneva, 1946.
———. *Creative Evolution*. Macmillan, London, 1911.
———. *Matter and Memory*. G. Allen and Unwin, London, 1911.
Berkeley, George. *Works*, ed. Jessop and Luce. Nelson, London, 1948.
Bilaniuk, O. M., Brown, S. L., De Witt, B., Necomb, W. A., Sachs, M., Sudarshan, E. C. G., Yoshikawa, S. "More about Tachyons," *Physics Today*, 22, no. 12 (1969).
Bilaniuk, O. M., Sudarshan, E. C. G. "Particles beyond the Light Barrier," *Physics Today*, May (1969).
Black, Max. *Philosophical Analysis*. Prentice Hall, Englewood Cliffs, NJ, 1963.

Blake, R. R., and Ramsey, G. V. (eds.). *Perception: An Approach to Personality*. Ronald Press Co., New York, 1951.

Blanshard, B. *The Nature of Thought*. George Allen and Unwin, London, 1939.

———. *Reason and Analysis*. Open Court, La Salle, IL, 1962.

Blum, H. *Time's Arrow and Evolution*. Princeton, NJ, 1955.

Boethius. *The Consolations of Philosophy*, trans. Richard Cooper. Bobbs-Merrill, Indianapolis, IN, 1962; trans. V. E. Watts, Penguin Books, Harmondsworth, 1969.

Bohm, David. *Fragmentation and Wholeness*. Van Leer Foundation Series, Jerusalem, 1976.

———. *Wholeness and the Implicate Order*. Routledge and Kegan Paul, London; Boston, 1980, 1983.

Bohm, D., and Aharanov, Y. "Discussion of Experimental Proof for the Paradox of Einstein, Podolsky and Rosen," *Physical Review*, 103 (1957).

Bohm, D., and Hiley, B. J. "Non-Relativistic Particle Systems," *Physics Reports*, 144, 6 (1987)

Bohm, D., and Krishnamurti, J. *The Ending of Time*. Harper and Row, San Francisco, CA, 1985.

Bohr, Niels. *Atomic Theory and the Description of Nature*. Cambridge University Press, Cambridge, 1934.

Bonner, J. T. *Morphogenesis*. Princeton University Press, Princeton, NJ, 1952.

———. *The Evolution of Development*. Cambridge University Press, Cambridge, 1958.

Borger, R., and Seaborne, A. E. M. *The Psychology of Learning*. Penguin Books, Harmondsworth, 1966.

Borresen, C. R., and Lichte, W. H., "Shape-Constancy: Dependence upon Stimulus Familiarity," *Journal of Experimental Psychology*, 63 (1962).

Bradley, F. H. *Appearance and Reality*. Clarendon Press, Oxford, 1897, 1930.

———. *Ethical Studies*. Clarendon Press, Oxford, 1927.

———. *The Principles of Logic*. Clarendon Press, Oxford, 1922.

Brain, Sir W. Russell. *Mind, Perception and Science*, 3 vols. Blackwell, Oxford, 1951.

Broad, C. D. *Examination of McTaggart's Philosophy*. Cambridge University Press 1933–38.

———. *Mind and Its Place in Nature*. Kegan Paul, Trench, London, Harcourt Brace, New York, 1925.

———. *Scientific Thought*. Kegan Paul, Trench, London, 1923; Humanities Press, NJ, 1952.

Brunner, Constantin. *Science, Spirit, Superstition*, trans. Abraham Suhl, ed. Walter Bernhard. G. Allen and Unwin, London, 1965.

Bruner, J. S., and Krech, D. *Perceptiona and Personality*. Duke University Press, Durham, NC, 1949.

Bruner, J. S., and Postman, L. "On the Perception of Incongruity," *Journal of Personality*, 18 (1949).

Bub, Jeffrey. "How to Kill Schroedinger's Cat," in *The World View of Contemporary Physics: Does It Need a New Metaphysics?* ed. R. Kitchener. State University of New York Press, Albany, NY, 1988.

Campbell, C. A. "The Mind's Involvement in 'Objects', An Essay in Idealist Epistemology," in *Theories of the Mind*, J. Scher, ed. Free Press of Glencoe, IL; Macmillan, London, 1962.

Capra, Fritjoff. *The Tao of Physics.* Wildwood House, London; Random House, New York, 1975, 1983.

———. "The Role of Physics in the Current Change of Paradigm," in *The World View of Contemporary Physics: Does It Need a New Metaphysics?* R. Kitchener, ed. State University of New York Press, Albany, NY, 1988.

Čapek, M. *Bergson and Modern Physics. A Reinterpretation and Reevaluation.* Reidel, Dordrecht, 1971.

———. *The Philosophical Impact of Contemporary Physics.* Van Nostrand, Princeton, NJ, 1961.

Chisholm, Roderick. "The Theory of Appearing," in *Philosophical Analysis*, ed. M. Black. Prentice Hall, Englewood Cliffs, NJ, 1963.

Churchland, P. M. *Matter and Consciousness.* Bradford Books, MIT Press, Cambridge, MA, 1984.

Cohen, R., and Wartoffsky, M., eds. *Hegel and the Sciences*, Boston Studies in the Philosophy of Science, Vol. 64. Reidel, Dordrecht; Boston and London, 1984.

Collingwood, R. G. *An Essay on Metaphysics.* Oxford University Press, Oxford, 1940.

———. *An Essay on Philosophical Method.* Oxford University Press, Oxford, 1934, 1965.

———. *The New Leviathan.* Oxford University Press, Oxford, 1942.

———. *The Idea of Nature.* Oxford University Press, Oxford, 1945.

———. *The Idea of History.* Oxford University Press, Oxford, 1946, 1956.

———. *Essays in the Philosophy of History*, ed. W. Debbins. University of Texas Press, Austin, TX, 1965.

Copernicus. *De Revolutionibus Orbium Coelestium.* Culture et Civilization, Brussels, 1966.

Cornford, F. *From Religion to Philosophy, A Study in the Origins of Western Speculation.* Arnold, London; Longmans Green, New York, 1912.

Craik, K. *The Nature of Explanation.* Cambridge University Press, Cambridge, 1952.

d'Abro, A. *The Evolution of Scientific Thought from Newton to Einstein.* Boni and Liveright, New York, 1927; 2nd. ed., revised and enlarged, Dover, New York, 1950.

———. *The Rise of the New Physics*, 2 vols. Dover, New York, 1951.

Darwin, Charles. *Autobiography*, ed. F. Darwin. Schumann, New York, 1950.

———. *The Voyage of the Beagle*, ed. L. Engel. Doubleday, New York, 1962.

———. *The Origin of Species.* Modern Library, New York, 1936.

Davies, Paul. *God and the New Physics.* Dent, London, 1983, Penguin Books, Harmondsworth, 1984, 1986.

———. *Other Worlds.* Dent, London, 1980.

deBroglie, L. *The Revolution in Physics: A Non-Mathematical Survey of Quanta*, trans. R. W. Wiemeyer. Routledge and Kegan Paul, London, 1954; Noonday Press, New York, 1953.

Descartes, René. *Discourse on Method, Meditations, Philosophical Works*, trans. Haldane and Ross. Cambridge University Press, Cambridge, 1931; reprinted 1975.

D'Espagnat, B. *The Conceptual Foundations of Quantum Mechanics.* Benjamin, Reading, MA, 1976.

DeWitt, B. S., and Graham, N. *The Many-Worlds Interpretation of Quantum Mechanics.* Princeton University Press, Princeton, NJ, 1973.

Dreyfus, H. *What Computers Can't Do.* Harper and Row, New York, 1972; revised ed. 1979.

Dreyfus, H., and Rabinow, Paul. *Michel Foucault; Beyond Structuralism and Hermeneutics.* University of Chicago Press, Chicago, IL, 1982, 1983.

Driesch, H. *Science and Philosophy of the Organism.* A. & C. Black, London, 1908.

———. *Mind and Body.* Methuen, London, 1927.

———. *The Problem of Individuality.* Macmillan, London, 1914.

Eccles, Sir J. C. *The Neurophysiological Basis of Mind.* Oxford University Press, Oxford, 1953.

Eddington, Sir Arthur. *The Nature of the Physical World.* Cambridge University Press, Cambridge, 1928.

———. *The Expanding Universe.* Cambridge University Press, Cambridge, 1933.

———. *New Pathways in Science.* Cambridge University Press, Cambridge, 1935.

———. *The Philosophy of Physical Science.* Cambridge University Press, Cambridge, 1939.

———. *Space, Time and Gravitation.* Cambridge University Press, Cambridge, 1950.

Einstein, A. *Relativity, the Special and General Theories,* trans. R. W. Lawson. Methuen, London, 1954.

———. *The Meaning of Relativity.* Methuen, London, 1956.

Einstein, A., and Infeld, L. *The Evolution of Physics.* Simon and Schuster, New York, 1954.

Einstein, Podolsky, and Rosen. "Can Quantum-Theory Description of Physical Reality Be Considered Complete?", *Physical Review,* 47 (1935).

Eliade, M. *From Prometheus to Zen: A Thematic Source Book of the History of Religions.* Harper and Row, New York, 1967.

Feyerabend, P. "Consolation for the Specialist," in *Criticism and the Growth of Knowledge,* ed. I. Lakatos and A. Musgrave. Cambridge University Press, Cambridge, 1970.

Feynman, R. P. *QED: The Strange Theory of Light and Matter.* Princeton University Press, Princeton, NJ, 1985.

———. *Surely You're Joking, Mr. Feynman!* Bantam, Books, New York, 1986.

Fichte, J. G. *Grundlage der Gesamten Wissenschaftslehre,* in *Sämmtliche Werke.* Mayer und Muller, Leipzig, 1845–46.

———. *Science of Knowledge,* trans. Peter Heath and John Lachs. Cambridge University Press, New York, 1982.

Freeman, G. L. *The Energetics of Human Behavior.* Cornell University Press, Ithaca, NY, 1948.

———. "The Problem of Set," *American Journal of Psychology,* 52 (1939).

Gale, G. "The Anthropic Principle," *Scientific American,* 154 (Dec. 1981).

Galilei, Galileo. *Dialogue Concerning the Two Chief World Systems,* Stillman Drake, trans. University of California Press, Berkeley, CA, 1962.

Gamow, George. *One, Two, Three . . . Infinity.* Bantam Books, New York, 1961.

Gelven, M. *Commentary on Heidegger's Being and Time.* Harper and Row, New York, Evanston, London, 1970.

Gibson, J. J. *Perception and the Visual World.* Houghton Miflin, Boston, 1950.

Gleick, J. *Chaos: Making a New Science.* Viking, Penguin Books, New York, and Harmondsworth, 1987.

Globus, G. G. "On 'I': The Conceptual Foundations of Responsibility," *American Journal of Psychiatry,* 137 (1980).

Gödel, K. "Über formal unentscheidbare Sätze der Principia Mathematica und Verwändter Systeme," *Monatshefte für Mathematik und Physik*, 38, 173–98. (1931). ————. *Collected Works*, ed. Solomon Feferman, et al. Oxford University Press, Oxford and New York, 1986.

Goetz, H. *Denken ist Leben*. Athenäum Verlag, Frankfurt-am-Main, 1987.

Goldschmidt, R. *The Material Basis of Evolution*. Yale University Press, New Haven, CT, 1940.

Gombrich, E. H. *Art and Illusion*. Phaidon Press, London, 1959, 1962; Bollingen, New York, 1961.

Greenberg, G. *The Symbiotic Universe: Life and Mind in the Cosmos*. Morrow and Co., Inc., New York, 1988.

Gribbin, John. *In Search of Schrödinger's Cat*. Bantam Books, New York, 1984. ————. *In Search of the Big Bang*. Bantam Books, New York, 1986.

Grier, Philip (ed.). *Dialectic and Contemporary Science*. University Press of America, Lanham, MD, 1989.

Haldane, J. S. *Organism and Environment*. The Terry Lectures. Yale University Press, New Haven, CT, 1917.

Hanson, N. R. *Patterns of Discovery*. Cambridge University Press, Cambridge, 1958.

Harris, E. E. *The Survival of Political Man*. Witwatersrand University Press, Johannesburg, 1950.

————. *Nature, Mind and Modern Science*. George Allen and Unwin, London, 1954.

————. "Time and Change," *Mind*, LXVI (1957).

————. "Teleology and Teleological Explanation," *Journal of Philosophy*, LVI, no. 1 (1959).

————. "Mind and Mechanical Models," in *Theories of the Mind*, ed. J. Scher. Free Press of Glencoe, Chicago, IL, 1962.

————. *The Foundations of Metaphysics in Science*. George Allen and Unwin, London, 1965, reprinted by The University Press of America, Lanham, MD, New York, London, 1983.

————. *The Reality of Time*. State University of New York Press, Albany, NY, 1988.

————. *Annihilation and Utopia*. George Allen and Unwin, London, 1966.

————. *Hypothesis and Perception, the Roots of Scientific Method*. George Allen and Unwin, London, 1970.

————. "Is the Real Rational?", *Contemporary American Philosophy*, ed. John E. Smith. George Allen and Unwin, London; Humanities Press, Atlantic Highlands, NJ, 1970.

————. "Epicyclic Popperism," *The British Journal for the Philosophy of Science*, 23 (1972).

————. "Dialectic and Scientific Method," *Idealistic Studies*, III, no. 1, 1973.

————. *Salvation from Despair, a Reappraisal of Spinoza's Philosophy*. Martinus Nijhoff, The Hague, 1973.

————. "Teleonomy and Mechanism," *Proceedings of the XVth International Congress of Philosophy*. 1973.

————. *Perceptual Assurance and the Reality of the World*. Clark University Press, New York, 1974.

————. *Atheism and Theism*. Tulane Studies. Martinus Nijhoff, The Hague, 1975.

————. "Coherence and Its Critics," *Idealistic Studies*, V, 3 (1975).

————. "Science, Metaphysics and Teleology," in *The Personal Universe*, ed.

Thomas E. Wren. Humanities Press, Atlantic Highlands, NJ, 1975.

———. "Time and Eternity," *Review of Metaphysics*, XXIX, 3 (1976).

———. "The Problem of Self-Constitution in Idealism and Phenomenology," *Idealistic Studies*, VII, 1 (1977).

———. "Blanshard on Perception and Free Ideas," in *The Philosophy of Brand Blanshard*, ed. P. Schilpp. Open Court, La Salle, IL, 1980.

———. *An Interpretation of the Logic of Hegel*. University Press of America, Lanham, MD, New York and London, 1983.

———. *Formal, Transcendental and Dialectical Thinking*. State University of New York Press, Albany, NY, 1987.

———. "Contemporary Physics and Dialectical Holism," in *The World View of Contemporary Physics: Does It Need a New Metaphysics?*, ed. R. Kitchener. State University of New York Press, Albany, NY, 1988.

———. "The Universe in the Light of Contemporary Scientific Developments," in *Bell's Theorem and Conceptions of the Universe*, ed. M. Kafatos. Kluwer Academic, Dordrecht, Boston, London, 1989.

Harrison, Edward. *Masks of the Universe*. Collier, Macmillan, New York and London, 1985.

Hawking, Stephen. *A Brief History of Time, from the Big Bang to Black Holes*. Bantam Books, Toronto, New York, London, 1988.

Hebb, D. O. *The Organization of Behaviour*. Wiley, New York, 1949.

Hegel, G. W. F. *Gesammelte Werke*, ed. F. Hogemann and W. Jaeschke. Felix Meiner Verlag, Hamburg, 1981—.

———. *Enzyklopädie der philosophischen Wissenschaften*, *Werke* (Vols VI–VIII). Suhrkamp Verlag, Frankfurt-am-Main, 1971–78.

———. *Encyclopaedia of Philosophical Sciences: Philosophy of Nature*, trans. A. V. Miller, foreword by J. N. Findlay. Clarendon Press, Oxford, 1975.

———. *Phänomenologie des Geistes*, *Werke* (Vol III). Suhrkamp Verlag, Frankfurt-am-Main, 1971–78.

———. *Phenomenology of Mind*, trans. J. Baillie. G. Allen and Unwin, London, 1910; revised 1931, reprinted, 1955, 1966.

———. *Phenomenology of Spirit*, trans. A. V. Miller. Oxford University Press, Oxford, 1977.

Heidegger, Martin. *Holzwege*, Frankfurt-am-Main, 1952.

———. *Einführung in der Metaphysik*. M. Wiemeyer, Tübingen, 1953; trans. Ralph Manheim as *Introduction to Metaphysics*, Yale University Press, New Haven, CT, 1959.

———. *Sein und Zeit*, M. Wiemeyer. Tübingen, 1967. Trans. J. Macquarrie and E. Robinson as *Being and Time*. S. C. M. Press, London, 1962.

Heisenberg, W. *Das Naturbild der heutigen Physik*. Rowolt, 1955. Trans. J Pomerans, *The Physicist's Conception of Nature*. Harcourt Brace, New York, 1958; Greenwood Press, Westport, CT, 1970.

———. *Philosophical Problems of Nuclear Science*. Faber & Faber, London, 1952.

———. *Physics and Philosophy: The Revolution in Modern Science*. Faber & Faber, London, 1952; Harper and Row, New York, 1959, 1962.

———. *Wandelungen in den Grundlagen der Naturwissenschaft*. S. Hirzel, Leipzig, 1945.

Hesse, M. B. *Forces and Fields*. Nelson, London, 1961.

Hirst, J. R. *Problems of Perception*. George Allen and Unwin, London, 1959.

Hobbes, Thomas. *Leviathan*, ed. M. Oakeshott. Clarendon Press, Oxford, 1946.

Hofstadter, D. R., *Gödel, Escher, Bach*. Vintage Books, New York, 1980.

Hofstadter, D. R., and Dennett, D. C. (eds.). *The Mind's I*. Basic Books, New York; Penguin Books, Harmondsworth, 1981.

Hogben, L. T. *The Nature of Living Matter*, Routledge and Kegan Paul, London, 1931.

Hubel, D. H. *Eye, Brain, and Vision*. Scientific American Library, Series 22. W. H. Freeman, New York, 1988.

Hume, David. *Treatise of Human Nature*, ed. L. A. Selby-Bigge. Oxford at the Clarendon Press; Oxford University Press, New York, 1978; Dutton, London, 1911.

————. *Dialogues Concerning Natural Religion* (1779). Bobbs-Merrill, Indianapolis, IN, 1981.

Husserl, E. *Die Krisis der europäischen Wissenschaften und die transcendentale Philosophie*, ed. E. Ströker. Felix Meiner Verlag, Hamburg, 1977.

————. *Erfahrung und Urteil*. Felix Meiner Verlag, Hamburg, 1972.

————. *Experience and Judgment*, trans. J. S. Churchill and K. Ameriks. Northwestern University Press, Evanston, IL, 1973.

————. *Husserliana. Gesammelte Werke*. Martinus Nijhoff, The Hague, 1950—.

————. *Formale und Transcendentale Logik. Husserliana*, Band XVIII. Martinus Nijhoff, The Hague, 1950—.

————. *Ideen zu einer Phänomenologie und phänomenologische Philosophie*. Martinus Nijhoff, The Hague, 1982.

————. *Logische Untersuchungen*. M. Niemeyer, Tübingen, 1968.

————. *Formal and Transcendental Logic*, trans. Dorian Cairns. Martinus Nijhoff, The Hague, 1969.

————. *Ideas*, trans. W. Boyce Gibson. George Allen and Unwin, London, 1931, 1956.

————. *Logical Investigations*, trans. John Findlay. Routledge and Kegan Paul, London, 1970.

————. *Phenomenology of Internal Time Consciousness*, ed. Martin Heidegger, trans. J. S. Churchill. Indiana University Press, Bloomington and London, 1964, 1973.

————. *The Crisis of the European Sciences*, trans. David Carr. Northwestern University Press, Evanston, IL, 1970.

Huxley, J. *Evolution, the Modern Synthesis*. Harper, London and New York, 1942.

————. *Evolution in Action*. Harper, New York, 1953.

Ittleson, H. W. "Size as a Cue to Distance: Static Localization," *American Journal of Psychology*, 64 (1951).

Ittleson, H. W., and Cantril, H. *Perception, a Transactional Approach*. Doubleday, Garden City, NY, 1954.

Ittleson, H. W., and Kilpatrick, F. P. "Experiments in Perception," *Scientific American*, 185, no. 2 (1951).

Jasper, H. H. (ed.). *The Reticular Formation of the Brain*. Henry Ford Hospital International Symposium, 1958.

Jeans, Sir James. *Physics and Philosophy*. Cambridge University Press, Cambridge, 1942.

Joachim, H. H. *Logical Studies*. Clarendon Press, Oxford, 1942.

Jonas, H. *The Phenomenon of Life*. Harper and Row, New York, 1966.

————. *Philosophical Essays*. Englewood Cliffs, NJ, 1974.

Kant, Immanuel. *Kritik der reinen Vernunft*, ed. G. Hartenstein. Leopold Voss, Leipzig, 1853; ed. Raymond Schmidt, Felix Meiner Verlag, Leipzig, 1926, 1930.

————. *Critique of Pure Reason*, trans. Kemp Smith. Macmillan, London, 1929.

Kafatos, Menas (ed.). *Bell's Theorem and Conceptions of the Universe*. Kluwer Academic Publishers, Dordrecht, Boston, London, 1989.

Kaku, Michio, and Trainer, Jennifer. *Beyond Einstein: The Cosmic Quest for the Theory of the Universe*. Bantam Books, Toronto, New York, London, 1987.

Katner, S. (ed.). *Vision and Action*. Rutgers University Press, New Brunswick, NJ, 1953.

Kitchener, Richard (ed.). *The World View of Contemporary Physics: Does It Need a New Metaphysics?* State University of New York Press, Albany, NY, 1988.

Koestler, A., and Smythies, J. R. *Beyond Reductionism*. Hutchinson, London, 1969.

Koffka, K. *The Growth of the Mind*. Kegan Paul, Trench, London, 1928.

———. *Principles of Gestalt Psychology*. Routledge and Kegan Paul, London; Harcourt Brace, New York, 1935.

Köhler, W. *The Mentality of Apes*. Harcourt Brace, New York, NY, 1925.

Kuhn, T. *The Structure of Scientific Revolutions*. University of Chicago Press, Chicago, IL, 1962, 1970.

Lakatos, I., and Musgrave, A. *Criticism and the Growth of Knowledge*. Cambridge University Press, Cambridge, 1970.

Langer, S. *Philosophical Sketches*. John Hopkins University Press, Baltimore, 1962.

———. *Mind: An Essay on Human Feeling*. 3 vols. John Hopkins University Press, Baltimore, 1967–83.

Lashley, K. S. *Brain Mechanisms and Intelligence*. University of Chicago Press, Chicago, IL, 1929.

Lavoisier, A. L. *Traité elementaire de Chimie*. Paris, 1789; Brussels, 1965.

Leclerc, I. *The Nature of Physical Existence*. George Allen and Unwin, London, 1972.

Levy-Strauss, C. *Le Pensée Sauvage*. Libraire Plon, Paris, 1962. Translated as *The Savage Mind*. George Weidenfeld and Nicholson, Ltd., London; University of Chicago Press, Chicago, IL, 1966, 1968.

Locke, John. *An Essay Concerning Human Understanding*, ed. Alexander Campbell Fraser. Clarendon Press, Oxford, 1894; Dover, New York, 1959.

London, I. "A Russian Report on Post-Operative Newly-Seeing," *American Journal of Psychology*, 73 (1960).

Lovelock, J. E. *Gaia, a New Look at Life on Earth*. Oxford University Press, Oxford, 1979.

MacPherson, Thomas. *The Argument from Design*. Macmillan, London, 1972.

McTaggart, J. E. *The Nature of Existence*. Cambridge University Press, Cambridge, 1927.

Mannheim, K. *Ideology and Utopia*. Harcourt Brace, New York, 1936; Kegan Paul, Trench, London, 1936, 1949.

Margenau, H. *The Nature of Physical Reality*. McGraw Hill, New York, 1950.

———. "The Exclusion Principle and its Philosophical Importance," *Philosophy of Science*, 11 (1944).

Medawar, Sir P. B. "Critical Notice on *The Phenomenon of Man* by Pierre Teilhard de Chardin," *Mind*, LXX, no. 277 (Jan. 1961).

———. *The Art of the Soluble*. Methuen, London, 1968.

Merleau-Ponty, M. *The Phenomenology of Perception*, trans. C. Smith. Routledge and Kegan Paul, London, 1962.

———. *The Structure of Behaviour*, trans. A. L. Fisher, Boston, 1963.

———. *The Primacy of Perception*. Northwestern University Press, Evanston, IL, 1964.

Metzger, W. "Optische Untersuchungen am Ganzfeld," *Psychologische Forshung*, 15 (1930).

Milne, E. A., *Relativity, Gravitation and World Structure*. Clarendon Press, Oxford, 1935.
————. "Fundamental Concepts of Natural Philosophy," *Proceedings of the Royal Society of Edinburgh*, Sec. A, Vol. 62 (1943–44), Part I.
————. *Kinematic Relativity*. Clarendon Press, Oxford, 1948.
Miller, I. *Husserl, Perception and Temporal Awareness*. M.I.T. Press, Cambridge, MA, 1984.
Monod, Jacques. *Le hazard et la necessité*. Paris, 1970.
————. *Chance and Necessity*, trans. A. Wainhouse. New York, 1970; Collins, London, 1972.
Morowitz, J. H. *Mayonnaise and the Origins of Life*. Scribners, New York, 1985.
Morgan, T. H. *The Scientific Basis of Evolution*. Faber & Faber, London, 1932.
Mure, G. R. *Aristotle*. Benn, London, 1932.
————. *Introduction to Hegel*. Oxford University Press, Oxford, 1940.
Nagel, E., and Newman, J. R. *Gödel's Proof*. Routledge and Kegan Paul, London, 1958.
Natsoulas, Thomas. "Conscious Perception and the Paradox of Blindsight," in *Aspects of Consciousness*, ed. G. Underwood. Academic Press, London, 1982.
Needham, J. *Biochemistry and Morphology*. Cambridge University Press, Cambridge, 1942.
————. *Time, the Refreshing River*. G. Allen and Unwin, London, 1944.
Newton, Sir Isaac. *Opticks*. Dover Publications, New York, 1952.
————. *Mathematical Principles of Natural Philosophy*, ed. F. Cajori, trans. Andrew Motte. University of California Press, Los Angeles, CA, 1966.
————. *Philosophiae Naturalis Principia Mathematica*, 2nd. ed., Alexander Koyré and Bernard Cohen, eds. Cambridge University Press, Cambridge, 1972.
Oparin, A. I. *The Origin of Life*. Oliver & Boyd, London, 1957.
Paley, William. *Natural Theology*. S. King, New York, 1824.
Park, D. *The Image of Eternity*. University of Massachusetts Press, Amherst, MA, 1980.
Paul, G. A. "Is There a Problem about Sense-Data?," in *Proceedings of the Aristotelian Society*, Supplementary Volume XV (1936).
Penrose, Roger. *The Emperor's New Mind: Concerning Computers, Minds, and the Laws of Physics*. Oxford University Press, Oxford, Melbourne, New York, 1989.
Piaget, J. *The Psychology of Intelligence*. Routledge and Kegan Paul, London, 1950.
————. *The Constitution of Reality in the Child*, trans. Margaret Cook. Basic Books, New York, 1954.
————. *Le Structuralisme*. Presses Universitaires de France, Paris, 1968. Trans. Chaninah Maschler as *Structuralism*. Basic Books, New York, 1970.
Planck, M. *The Philosophy of Physics*. G. Allen and Unwin, London, 1936; Norton & Co., New York, 1936.
————. *The Universe in the Light of Modern Physics*. G. Allen and Unwin, London, 1937; Norton & Co., New York, 1951.
Plato. *Dialogues of Plato*, trans. Benjamin Jowett. Clarendon Press, Oxford, 1953.
————. *The Works of Plato*, ed. Edith Hamilton and Huntington Cairns. Princeton, 1963.
————. *Plato's Theory of Knowledge; Theaetetus and Sophist*, trans. F. Cornford. Routledge and Kegan Paul, London, 1935.
————. *Plato's Cosmology; Timaeus*, trans. F. Cornford. Routledge and Kegan Paul, London, 1937, 1957.
————. *Republic*, trans. F. Cornford. Oxford University Press, Oxford, 1945.

Poincaré, H. *Science and Hypothesis*. Dover Publications, New York, 1952.
Polanyi, M. *Personal Knowledge*. Routledge and Kegan Paul, London, 1958.
Popper, Sir Karl. *Conjectures and Refutations*. Routledge and Kegan Paul, London, 1965; Basic Books, New York, London, 1962, 1963.
──────. *Objective Knowledge; An Evolutionary Approach*. Clarendon Press, Oxford, 1979.
──────. *The Logic of Scientific Discovery*. Hutchinson, London, 1959.
──────. *The Open Society and Its Enemies*. Princeton University Press, Princeton, NJ; London, 1950, 1966.
Price, H. H. *Perception*. Methuen, London, 1950.
──────. "Appearing and Appearances," *American Philosophical Quarterly*, I (1964).
Prigogine, I. "Irreversibility as Symmetry-Breaking Process," *Nature*, 246 no. 67 (1973).
──────. "The Rediscovery of Time," in *The World View of Contemporary Physics: Does It Need a New Metaphysics?*, ed. R. Kitchener. State University of New York Press, Albany, NY, 1988.
Prigogine, I., and Nicolis, G. *Self-Organization in Non-Equilibrium Systems*. Wiley, New York, 1977.
Quinton, A. M. "The Problem of Perception," *Mind*, LXIV (1955).
──────. "Perception and Thinking," *Proceedings of the Aristotelian Society*, Supplementary Vol. XLII (1968).
Richter, D. (ed.). *Perspectives in Neuropsychiatry*. H. K. Lewis, London, 1950.
Riesen, A. H. "The Development of Visual Perception in Man and Chimpanzee," *Science*, 106 (1947).
Rohrer, J. H., and Sherif, M. (eds.). *Social Psychology at the Cross-Roads*. Harper, New York, 1951.
Ross, Sir David. *Aristotle*. Methuen, London, 1937, 1949.
Russell, Bertrand. *Human Knowledge, Its Scope and Limits*. George Allen and Unwin, London, 1948.
──────. "Logical Atomism," *Contemporary British Philosophy*, Series I. George Allen and Unwin, London, 1924.
──────. *Our Knowledge of the External World*. Norton, New York, 1929.
──────. *The Philosophy of Logical Atomism*. University of Minnesota Press, Minneapolis, MN, 1959.
──────. *The Principles of Mathematics*. G. Allen and Unwin, London, 1937.
Russell, E. S. *The Directiveness of Organic Activities*. Cambridge University Press, Cambridge, 1945.
Ryle, Gilbert. *The Concept of Mind*. Hutchinson, London, 1949.
──────. "Sensation," in *Contemporary British Philosophy*, Third Series. George Allen and Unwin, London, 1956.
Salleri, F. (ed.). *Quantum Mechanics versus Local Realism: The Einstein, Podolsky and Rosen Paradox*. Plenum Press, New York, 1987.
Schelling F. W. J. *System des Transcendentalen Idealismus*. Klett Cotta, Tübingen, 1800; Felix Meiner Verlag, Leipzig, 1957.
──────. *System of Transcendental Idealism*, trans. Peter Heath. University of Virginia Press, Charlottesville, VA, 1978.
──────. *Ideen zu eine Philosophie der Natur*, trans. as *Ideas for a Philosophy of Nature* by Errol Harris and Peter Heath. Cambridge University Press, Cambridge and New York, 1988.
Scher, J. (ed.). *Theories of the Mind*. Free Press of Glencoe, New York, 1962.

Schilpp, P. A. (ed.). *The Philosophy of A. N. Whitehead*. Northwestern University Press, Evanston IL, 1941.
———. *Albert Einstein, Philosopher, Scientist*. Northwestern University Press, Evanston, IL, 1949.
———. *The Philosophy of Brand Blanshard*. Open Court, Lasalle, IL, 1980.
Schrödinger, E. *Space-Time Structure*. Cambridge University Press, Cambridge, 1950.
———. *Science and Humanism*. Cambridge University Press, Cambridge, 1952.
———. *What Is Life? and Other Scientific Essays*. Doubleday, New York, 1956.
Schwartz, John. *Superstrings*. World Scientific, Singapore, 1985.
Sciama, D. W. *The Unity of the Universe*. Doubleday, Garden City, NY, 1959, 1961; Faber and Faber, London, 1959.
Sheldrake, R. *A New Science of Life*. Paladin, London, 1987.
Sherrington, Sir Charles. *Man on His Nature*. Cambridge University Press, Cambridge, 1951.
Sinnott, E. W. *The Problem of Organic Form*. Yale University Press, New Haven, CT, 1963.
Snell, B. *The Discovery of the Mind, the Greek Origins of European Thought*. Blackwell, Oxford, 1953.
Smith, John E. (ed.). *Contemporary American Philosophy*. George Allen and Unwin, London; Humanities Press, Atlantic Highlands, NJ, 1970.
Spinoza B. *Collected Works*, trans. and ed. E. Curley. Princeton University Press, Princeton, NJ, 1985—.
———. *Ethics*, trans. A Boyle. M. Dent, London, 1910.
———. *Ethics, and Selected Letters*, trans. S. Shirley. Hackett, Indianapolis, IN 1982.
———. *Opera quotquot reperta sunt*, ed. J. van Vloten and J. P. N. Land. Martinus Nijhoff, The Hague, 1913–14.
Stapp, H. P. "Are Faster-Than-Light Influences Necessary," in *Quantum Mechanics versus Local Realism: The Einstein, Podolsky and Rosen Paradox*, ed. F. Salleri. Plenum Press, New York, 1987.
———. "Quantum Mechanics and the Physicist's Conception of Nature: Philosophical Implications of Bell's Theorem," *The World View of Contemporary Physics: Does It Need a New Metaphysics?*, ed. R. Kitchener. State University of New York Press, Albany, NY, 1988.
Stebbing, Susan. *Philosophy and the Physicists*. Methuen, London, 1937.
Teilhard de Chardin, P. *Le phénoméne humain*. Editions du Seuil, Paris, 1955.
———. *The Phenomenon of Man*, trans. B. Wall, intro. by Julian Huxley, Collins, London; Harper and Row, New York, 1959.
Tenant, F. R. *Philosophical Theology*. Cambridge University Press, Cambridge, 1928, 1969.
Thirring, W. "Urbausteine der Materie," *Almanach der Österreichischen Akademie der Wissenschaften*, 118 (1968) (Imp. 1969).
Thom, R. *Stability and Morphogenesis*. Benjamin, Reading, MA, 1975.
Thomas, L. *The Lives of a Cell*. Viking Press, New York, 1974; Penguin Books, Harmondsworth, 1987.
Thorpe. H. *Learning and Instinct in Animals*. Methuen, London, 1956, 1963, 1964, 1966.
Thouless, R. H. "Phenomenal Regression to the Real Object," *British Journal of Psychology*, 21 (1931).
Tinbergen, N. *A Study of Instinct*. Oxford University Press, Oxford, 1952.

Verene, D. P. (ed.). *Hegel's Social and Political Thought.* Humanities Press, Atlantic Highlands, NJ, 1980.

Vernon, M. D. (ed.). *Experiments in Visual Perception.* Penguin Books, Harmondsworth, 1966.

————. *The Psychology of Perception.* Penguin Books, Harmondsworth, 1969.

von Bertalanffy, L. *Modern Theories of Development.* Oxford University Press, London, 1933.

von Neumann, J. *Mathematical Foundations of Quantum Mechanics.* Princeton University Press, Princeton, NJ, 1955.

von Senden, M. *Raum und Gestalt auffassung bei operierten Blindgeborenen vor und nach der Operation.* Leipzig, 1932.

————. *Space and Sight.* Free Press of Glencoe, Chicago, IL, 1960.

Waddington, C. H. *The Strategy of the Genes.* George Allen and Unwin, London, 1957.

Weinberg, S. *1979 Nobel Prize Lecture.* Nobel Foundation, Stockholm, 1980.

————. *The First Three Minutes: A Modern View of the Origin of the Universe.* Basic Books, 1977; Bantam Books, New York, 1984.

————. *Gravitation and Cosmology: Principles and Applications of the General Theory of Relativity.* Wiley, New York, 1972.

Weiskrantz, L. "Neuropsychology and the Nature of Consciousness," in *Mindwaves,* ed. C. Blakemore and S. Greenfield. Blackwell, Oxford, 1987.

Werner, H., and Wapener, S. "Sensory Tonic Field Theory of Perception," in *Perception and Personality,* ed. J. S. Bruner and D. Krech. Duke University Press, Duhram, NC, 1949.

Wheeler, J. A., and Zurek, W. H. (eds.). *Quantum Theory and Measurement.* Princeton University Press, Princeton, NJ, 1983.

Whitehead, A. N. *Adventures of Ideas.* Cambridge University Press, Cambridge; Macmillan, New York, 1933.

————. *An Enquiry into the Principles of Natural Knowledge.* Cambridge University Press, Cambridge, 1919, 1925; Dover, New York, 1982.

————. *Process and Reality.* Cambridge University Press, Cambridge, 1929. D. R. Griffin and D. W. Sherburn, eds. Free Press, New York, 1978.

————. *Science and the Modern World.* Cambridge University Press, Cambridge, 1926; Macmillan, New York, 1925, 1948; New American Library, New York, 1953.

————. *The Concept of Nature.* Cambridge University Press, Cambridge, 1920, 1971.

Whitrow, J. G. "On Synthetic Aspects of Mathematics," *Philosophy,* XXV, no. 95 (1950).

————. *The Natural Philosophy of Time.* Nelson, London and Edinburgh, 1961; Harper Torchbooks, New York and Evanston, 1963.

Whittaker, Sir Edmund. *From Euclid to Eddington,* Cambridge University Press, Cambridge, 1949.

Whittaker, R. H. *Communities and Eco-Systems.* Collier-Macmillan, New York, 1975.

Whyte, L. L. *The Unitary Principle in Physics and Biology.* Cresset Press, London, 1949.

Wigner, E. "Epistemology in Quantum Mechanics," in *Contemporary Physics: Triest Symposium,* 1968, Vol. II, pp. 431–438. International Atomic Energy Authority, Vienna, 1969.

Will, Clifford M. *Was Einstein Right? Putting General Relativity to the Test.* Basic Books, New York, 1986.

————. "Gravitational Theory," *Scientific American*, 24 (Nov. 1974).
Williams, D. *The Ground of Induction*. Harvard University Press, Cambridge, MA, 1947.
Wittgenstein L. *Tractatus Logico-Philosophicus*. Routledge and Kegan Paul, London, 1960.
————. *Notebooks, 1914/1916*. Blackwell, Oxford, 1961.
————. *Philosophical Investigations*. Routledge and Kegan Paul, London, 1967.
————. *On Certainty*. Blackwell, Oxford, 1969.
Wren Thomas E. (ed.). *The Personal Universe: Essays in Honor of John Macmurray*. Humanities Press, Atlantic Highlands, NJ, 1975.

Index

Boldface numbers indicate pages where a topic is treated more fully.